D1070812

The Faith of Biology
& the Biology of Faith

Columbia Series in Science and Religion

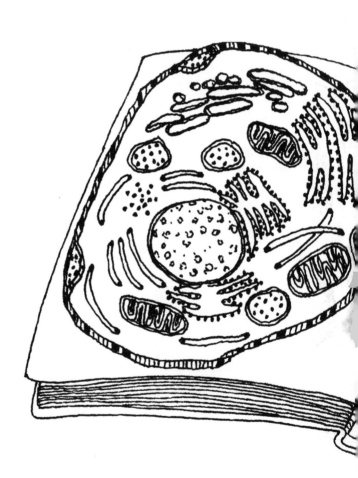

COLUMBIA UNIVERSITY PRESS   NEW YORK

Robert Pollack **The Faith of Biology
& the Biology of Faith**

Order, Meaning, and Free Will
in Modern Medical Science

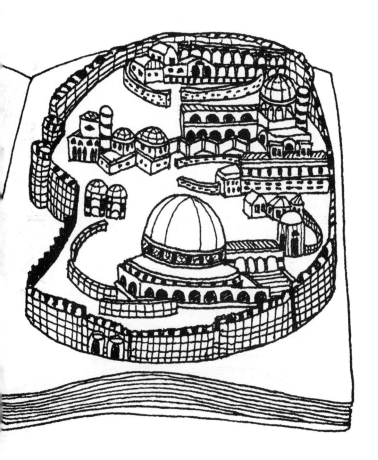

Columbia University Press

Publishers Since 1893

New York   Chichester, West Sussex

Copyright © 2000 Columbia University Press

Library of Congress Cataloging-in-Publication Data

Pollack, Robert

   The faith of biology & the biology of faith : order,
meaning, and free will in modern medical science / Robert
Pollack.

     p. cm. — (Columbia series in science and religion)

     Includes bibliographical references and index.

     ISBN 0-231-11506-7 (cloth : alk. paper) —
ISBN 0-231-11507-5 (pbk. : alk. paper)

     1. Judaism and science. 2. Meaning (Philosophy) — Re-
ligious aspects. 3. Natural selection. 4. Free will and determinism. 5. Genetics — Moral and ethical aspects. I. Title:
Faith of biology and the biology of faith. II. Title. III.
Series.

     BM538.S3 P65 2000

     296.3'75 — dc21   00-055594

Casebound editions of Columbia University Press books
are printed on permanent and durable acid-free paper.

Printed in the United States of America

c 10 9 8 7 6 5 4 3 2 1

Title page: Jerusalem and a cell are both busy places.
Jerusalem's Old City and the cell's nucleus respectively
codify and direct the comings and goings of people and
molecules.

Title page illustration by Amy Pollack
Designed by Lisa Hamm

Columbia Series in Science and Religion

University Seminars

Leonard Hastings Schoff Memorial Lectures

THE UNIVERSITY SEMINARS at Columbia University sponsor an annual series of lectures, with the support of the Leonard Hastings Schoff and Suzanne Levick Schoff Memorial Fund. A member of the Columbia faculty is invited to deliver before a general audience three lectures on a topic of his or her choosing. Columbia University Press publishes the lectures.

1993
David Cannadine, *The Rise and Fall of Class in Britain*

1994
Charles Larmore, *The Romantic Legacy*

1995
Saskia Sassen, *Sovereignty Transformed — States and the New Transnational Actors*

2000
Robert Pollack, *The Faith of Biology and the Biology of Faith: Order, Meaning, and Free Will in Modern Medical Science*

For Amy

Contents

Preface

> The human psyche can tolerate a great deal of prospective
> misery, but it cannot bear the thought that the future is be-
> yond all power of anticipation.
>
> — Robert Heilbronner, quoted in Wolpert,
> *The Unnatural Nature of Science*

W HEN I ARRIVED on the Columbia University campus in the fall of 1957, I had in my hands an acceptance letter with a light blue postage stamp picturing the dome of Low Memorial Library surrounded by "Man's Right to Knowledge and the Free Use Thereof," the motto of Columbia's 1954 bicentennial. The implicit promise made to me then by that stamp — that this place would give me knowledge and teach me how to make free use of it — was met by the odd lot of courses, conversation, and lab work that made up the life of an undergraduate physics major at that time. Indeed, the gift and the burden of Columbia's education — lots of facts and the power to argue against them — extended well past the physics and math I studied so intently, informing my adult intellectual life as a molecular biologist, writer, and Jew.

In early 1999, as I approached a half-century of residence at Columbia wondering what could possibly top my old motto for our upcoming quarter-millennial celebrations, I was invited to make a new and unexpected use of the original one when the University Seminars invited me to give the 1999 Schoff Memorial Lectures, a series of three public lectures. They could be on any topic I chose, providing only that it would be approached from more than one academic perspective and that it would be comprehensible to the widest possible university audience. In other words, I was being asked to show what I had done with my "right to knowledge and the free use thereof," and a recitation from — or of — my curriculum vitae would not serve the purpose.

I turned instead to a class of question that had been bothering me for some time, the ones arising at the junction of scientific and religious conceptions of the world. When these lectures allowed me to ask why such questions were not yet part of the academic life here at Columbia, that made me wonder, in turn, whether a avowedly secular place like Columbia could allow such questions to be asked in the proper spirit of free inquiry, with mutual respect for strong feelings and for the benefit of both worlds, or whether such an effort would be doomed to collapse in the mutual disregard so typical of failed interdisciplinary enterprises.

Happily, the administration of Columbia University agreed to let me find out. Since the fall of 1999 — as the last academic year of the millennium began — I have been the director of a new interdisciplinary part of the university, the Columbia Center for the Study of Science and Religion (CSSR). The center's programs include curricula, seminars, and public lectures and forums. The Schoff Lectures formed the first public program of the new center, and this book is their publishable precipitate. *The Faith of Biology and the Biology of Faith* is the first of a series of books on science and religion that the Center for the Study of Science and Religion will sponsor and Columbia University Press will publish in the coming years. That series will, I trust, be a fitting return on Columbia's investment in me some fifty years ago.

Acknowledgments

A professor is one who talks in someone else's sleep.
— W. H. Auden

I
T IS A PLEASURE to acknowledge the people who have helped me make proper use of the opportunity to give the 1999 Schoff Memorial Lectures and to prepare this book from them. No book is written in isolation; my first thanks go, as always, to my family: Amy, Marya, Mark, Barry.

It was this book's good fortune and my own to have had the helpful, enthusiastic support of many friends and colleagues, first among them my editor at Columbia University Press, Holly Hodder. Rita Charon, William Theodore de Bary, and Robert Glick — three distinguished colleagues and dear friends whose range of experience and skill maps some of Columbia's most interesting terrain — were kind enough to read my manuscript in advance and then nevertheless provide me with an introduction at each lecture.

Richard Deckelbaum, Herman Wouk, and Philip Kitcher were willing to go to the trouble of reading the manuscript after these lectures were presented but in time for me to fix many errors and fill the many gaps. For earlier gifts of time and help I am also alphabetically indebted to Kate Brauman, MargyRuth and Perry Davis, Neil Gillman, Liz Haberfeld, Bill Lebeau, John Loike, Shannon Robinson, Joanna Samuels, Adin Steinsaltz, Joe Thornton, David Weiss-Halivny, Iddo Wernick, and Rebecca Zausmer.

For the chance to even think of giving the Schoff Lectures and then writing this book, thanks go as well to colleagues at Columbia who gave me permission, encouragement, funds, and freedom: Dean Aaron Warner of the University Seminars, President George Rupp, Provost Jonathan Cole, Professor Jim Manley, Kate Wittenberg, Emily Lloyd, Mike Crow, Deborah Mowshowitz, Charles Sheer, Earle Kazis, Joanne Ben-Avi, Philip Milstein, and Punch Sulzberger. I am especially indebted to Robert Kraft, who arrived with me in the late 1950s and has since given me, and Columbia, so much by way of encouragement and support.

New York City and Chelsea, Vermont
Winter 1999–2000

The Faith of Biology
& the Biology of Faith

Introduction

He was ceasing to care for knowledge — he had no ambition for practice — unless they could both be gathered up into one current with his emotions; and he dreaded, as if it were a dwelling-place of lost souls, that dead anatomy of culture which turns the universe into a mere ceaseless answer to queries, and knows, not everything, but everything else about everything — as if one should be ignorant of nothing concerning the scent of violets except the scent itself for which one had no nostril.

— George Eliot, *Daniel Deronda*

GIVEN MY PERSONAL pick of any topics at the boundary of a religion and a science to choose for this book, I quickly eliminated all but two as being either too easy or too difficult. Mortality and natural selection were the remaining possibilities. While mortality had the wider audience, natural selection won on grounds of personal interest. As I prepared the Schoff Memorial Lectures, they prepared me for the obvious outcome of such a choice: I may have decided to begin with natural selection, but I could not end without touching on mortality as well. The bridge that I found myself building from one to the other was not made of the usual stuff of academic argument — facts, models, tests, logic, rationality — but of spontaneously upwelling feelings and intuitions, and of their unknowable origins. This was to be, in other words, a book about my religion as well as my science.

My Religion and Its Place in This Book

I am not a representative of religion as such, nor of science as such. I am a particular person with a history of experience in one religion and in one science: a Jewish molecular biologist. The example of evolution is interesting to me neither because of its capacity to generate controversy nor because most religions, including Judaism, continue to have great difficulty in absorbing the detailed facts and implications of the record of natural selection written in every corner of the world, from the DNA of our own cells to the fossils of our ancestors deep within the earth.

I picked it because evolution through natural selection explains certain facts of life that touch on matters of meaning and purpose and because the vision of the natural world these explanations produce is simply too terrifying and depressing to me to be borne without the emotional buffer of my own religion. This buffer is simple to describe: a Jewish understanding of our appearance by evolution through natural selection introduces an irrational certainty of meaning and purpose to a set of data that otherwise show no sign of supporting any meaning to our lives on earth beyond that of being numbers in a cosmic lottery with no paymaster.

I have chosen to write from this perspective — as I have chosen to take my religion seriously — in response to unbidden, spontaneous, strong inward feelings, feelings I am no longer able to keep from acknowledging. There is a price to be paid for approaching the data of science from such a perspective: I must write from the heart and place feelings on a par with facts, something a scientist is ordinarily obliged to avoid. In so doing, I am at risk of alienating an audience most important to me, the reasonable people — some of them my closest colleagues — who share my conclusions but not my belief in the existence of a caring God. To forestall that, let me agree with them here, before I begin the book, that matters of personal belief cannot finally be tested by science and that therefore I neither may nor shall pass judgment on anyone else's religion nor lack thereof.

More than that, I wish to agree in advance with an even deeper critique of what I am about to do: I acknowledge there is a wholly consistent alternative description of the natural world and our place in it that can lead one to exactly the actions I wish to encourage, all without any belief in God. I have colleagues and friends who hold this conviction, and it would be falsely naive of me to overlook their position, which is, in brief: The world has no intrinsic meaning or purpose; we are mortal; we wish it were not so; it is; let's make the best of it by our own human lights; what is wrong with that?

Nothing is wrong with that position. It used to be my own, but as I have gotten older I find I no longer can honestly hold to it. When I asked my teacher, Rabbi Adin Steinsaltz, how to respond to this criticism of the book by nonbelieving friends, he said, "If you know someone who says the Throne of God is empty, and lives with that, then you should cling to that person as a good, strong friend. But be careful: almost everyone who says that has already placed something or someone else on that Throne, usually themselves."

And there is the reason for this book: I find myself accepting the God of my ancestors not because I have so completely transcended the absurdity of seeking evidence to prove the unprovable — I doubt I will ever be happy arguing for the irrational — nor even because I am filled with revelatory certainty, but for the lesser, negative reason that if I do not make these arguments I will most certainly put some person, some ideology, some dream of completed science in God's place, and that is what I wish never again to do.

Denial, Rejection, and Acceptance

Any attempt to look at aspects of the natural world with an eye toward vesting them with meaning and purpose — let alone doing so from a religious perspective — has a very high probability of failure in the view of scientists and of triviality in the view of people for whom faith is its own answer. What choices are available if one is to escape both failure and triviality? Denial, rejection, and acceptance have all been tried by others, but, of these, only acceptance makes sense for my purposes.

Denial is the position taken by someone who will say that there is no meaning to the problem of meaning. Many scientists and many people of faith take this position, sounding eerily alike in their blithe disregard of the real pain felt by other, equally learned or equally pious peers. Denial is a habit of mind that does not fit with any model of the mind that depends solely on rationality; it is a wholly nonrational way to live with unwanted thoughts and feelings. It is the not-hearing, the not-understanding, the not-getting-it strategy that leaves the person who raises the problem of meaning in the natural world to imagine himself burdened with a wholly avoidable mental quirk. The benefit of denial is the avoidance of confrontation with the problem of meaning; its double price is the failure of honesty, and of empathy.

Rejection of the possibility that science and religion may be able to inform each other's visions of the world is a far more direct and rational approach than denial, for either a scientist or a religious person. A religious person who rejects data, whether because the facts would be too hard to bear if they were true or because they would contradict the obligations of a faith, has accepted a much larger field of experience than the person in denial and is therefore far easier to be with.

Such a person might reject evidence that evolution is driven by random mutations in DNA, because the consequent inperfectibility of all species including our own would be too hard to bear if it were so, and would therefore argue that we must simply be unable to see the patterns of regularity in preadaptive mutation that have driven evolution along a path to perfectibility. Such a person might also assume that death cannot be entropically irreversible since something — the soul — must survive death or that self-consciousness — the voice one hears in one's head when there is no one else in the room — cannot be simply the result of differential gene expression in the cells of the central nervous system, because, then, who are we? Or such a person might say that the notion of a God outside the constraints of time and space cannot be meaningful, because to vest such a notion with meaning would be to accept the unbearable possibility that a fraction of what exists may be not only unknown but unknowable.

Acceptance — rather than nonrational denial or rational rejection — is the simplest way to deal with problems at the intersect of a science and a religion. Acceptance requires no prior assumptions. Denying nothing of human experience, it accepts the equal validity of feelings and facts. In the case of evolution it means accepting the facts of biology and the wish to reject some of those facts, as equally real in both the scientific and religious senses of the word. By accepting both facts and the feelings, both can be acknowledged in scientific terms as hard-won details of the natural world and in religious terms as examples of an unknowable design capable of lending meaning and purpose to the facts. In this frame of mind the living world as described by science can be vested with meanings that emerge from a religious tradition, without rejection or denial of either the facts or the feelings involved.

Scientific Discourse and the Names of God

The crystallization of a religious understanding of the source of the set of unbidden feelings and revelatory ideas is the small but powerful word *God*. In my religion there is a deep and ancient reticence to dare to put a name on what is for us the essence of the unknowable. The first name for God in our revelatory texts, the name of the creator in Genesis, is *Elohim*. The subsequent four-lettered Name of God mistranslated in English as *Jehovah* cannot be pronounced meaningfully. The Jewish practice is to over-read this name with one or another euphemism. Of these euphemisms *Adonai*, or *Lord*, is perhaps most common, and *Ha-Shem*, or the Name, is perhaps purest.

Though the word *God* will not always do for us here, neither can I use any of these alternatives, because none of them will have meaning for many people I wish to engage in this discussion of the possibility of finding meaning in the orderliness of the living world. Using any name of God simply licenses otherwise-interested people to set my ideas aside in advance as having no content except for other people for whom the word *God* has meaning. There is another, different problem with saying *God* in this book, precisely for people for whom it does have meaning. There are at least as many religions — each with their own names for their conceptions of the unknowable — as there are sciences.

To use *God*, or any personally comfortable name of God, is to risk the alienation and loss of scientists and nonscientists whose religion is not my own. Yet, to talk with specificity and conviction about unknowability and science, I really have no choice but to talk largely about my religion as well as my science, and, to do that, I must use words that have force for me. Here is how I will proceed.

In some places I will be using the word *God* because that is what is meaningful to me. But because this may exclude someone for whom the unknowable is real and questions of personal meaning critical, even though no organized religion currently serves as a guide, I will use *unknowable* as an alternative, either as an adjective or as a noun. The adjective is unexceptionable, but the noun will look and sound odd. Nevertheless, by referring to the unknowable as an aspect of reality in this way, I am confident at least that I will not be giving anyone license to ignore what I say for semantic reasons.

The Argument of This Book

The first chapter of this book, based on the first of my three Schoff Lectures, is about meaning and purpose and the difficulty of finding either in the data of any science. Natural selection seems quite free of design, purpose, or direction, yet it has the potential to be a complete explanation of our presence as a species. Already science based on the facts of natural selection is able to explain a good deal about our minds, our brains, and our behaviors and claims to be able to explain our uses of language to teach and learn ideas, all without calling upon any natural intention or design or Designer. Even our most precious ideas — the idea of God among them — may arguably be no more than the product of natural selection for competing notions, so that the idea of purpose and meaning may have no meaning or purpose beyond their self-propagation as ideas. In this part of the book I attempt to save meaning from this pit by a close look at free will, the uniquely dangerous but powerful human capacity to choose a belief or an action on emotional grounds, rather than deciding it on the basis of rational argument.

The second chapter, based on the second lecture, considers the consequences of a loss of free will. The free choice of a scientist to accept an insight of science and pursue it by experimentation is not so different from the free choice of a religious person to accept the insight of a prophet and pursue it by emotional commitment. Both are examples of free will providing a way for meaning to emerge from the otherwise meaningless orderliness of the discoverable natural world. In both worlds the loss of free will produces passivity, dogma, and fundamentalism.

Free will need not be freely surrendered to be lost; sometimes it may simply be taken away. Illness is the most common way a person becomes deprived of her ability to make choices; when medicine works on behalf of a patient, the patient's free will is enhanced. In this context some basic medical science works in a patient's interest and some works in opposition to it.

The final chapter of the book, based on the third lecture, considers genetic medicine. This branch of basic medical science makes the promise that it will be able to forestall a particular illness one day in the future, but only if people today — including people not yet suffering any symptoms — agree in advance to learn their genetic fates. The imposition of this knowledge in the absence of treatment is the permanent loss of an essential component of a person's free will. In requiring this bargain genetic medicine risks reducing its scientific discoveries to purposeless and meaninglessness.

Genetic diagnosis of future illness need not be held in abeyance until treatments are available for the problems it can predict. It can be a safe tool for medicine even before then, providing that both doctor and patient realize — and act toward each other with respect for the fact — that in their mortality they are mutually obliged to protect each other's freedom to choose their fate.

chapter one **Order Versus Meaning:
Science and Religion**

> Science is meaningless because it gives no answer to our
> question, the only question important for us: what shall we
> do and how shall we do it?
> — Leo Tolstoy, quoted in Lewis Wolpert,
> *The Unnatural Nature of Science*

T HE SEAL OF Columbia College —
subsequently Columbia University — is almost a quarter of a millennium old. It personifies all of us,
faculty and students alike, as naked babies. Seated before us is
the ideal Teacher, the spiritual mother of us all, Alma Mater,
arms out, scepter of wisdom in her hand. Below her is a reference to chapter 2 of the first Epistle of Peter: "Wherefore lay-
ing aside all malice, and all guile, and hypocrisies, and envies
and all evil speakings / As new born babes, desire the sincere
milk of the word, that ye may grow thereby." Around her
shoulders is a fragment of a line in Latin from the Hebrew
Scriptures, psalm 36, line 10: "By Your light do we see light."
Together these Biblical references are a brilliant and poetic
evocation of the acts of teaching and learning.

I have been saying the words of the psalm in Hebrew every
day for a few years, since I found myself convinced of the
need to accept the Jewish obligation — kept by my ancestors

for thousands of years — to say traditional prayers every morning. I had become accustomed enough to the Hebrew some time ago to think about what I was saying as I said it. One morning it came to me with great force that the Latin of the seal's psalm was an edited version of the psalmist's intention. From the prayer I saw that the full line is "With You is the fountain of life; by Your light do we see light." And when I went back to look at the seal embedded at the threshold of Low Memorial Library, there indeed was the Hebrew for "Light of God" on a scroll in the Teacher's hand.

I wish I knew who designed our seal, and when, and why we leave out the first premise — that there is an unknowable Deity at the source of everything to be taught and everything to be learned, that everything known to be, and everything yet to be known, is surrounded by the Unknowable.

The Known, the Unknown, and the Unknowable

This book is about the boundary of the knowable and the unknowable. Science works at the boundary of the known and the unknown, a different place entirely. The unknowable as a notion does not come easily to the scientifically minded. Dealing with it is a project full of paradox, requiring that one talk about the unutterable and anatomize the unmeasurable. I chose to work at this new boundary, nevertheless, because I have the habits of thought of a scientist. As soon as the notion of the unknowable as distinct from the unknown placed itself before me, the shock changed both my career and the way I see the world. The unknown was the raw material of my career, and the notion that it might be bounded in this way seemed to me deeply subversive of the entire enterprise.

My first reaction was as a scientist: I kept this idea to myself and went on about my business as a laboratory director while I thought about how it might be put to a test of some sort. But then, like the spotted Dalmatian who leaps to run after any truck that sounds like a fire engine, my training — begun as a physics major in Columbia College — eventually obliged me to grab onto what was most interesting rather than what was expedient, to try to understand the notion of the unknowable in all its untestability, and to make what I could understand about the unknowable understandable to others in turn. That is what I have wished to do in this book, as I would have done for the data of my laboratory if that is where my curiosity had led me instead.

Science proceeds by the testing of hypotheses, but because a hypothesis that can stand up to testing expands the territory of the known, scientific hypotheses about the unknowable are not meaningful. Put another way, it is not worth a moment of anyone's time to seek the proof through science of any religious belief. And as this book is about the consequences of potential unknowability — a notion as foreign to many reasonable nonscientists as it is to the scientific method — I needed first to provide some working terminology for the unknowable, without calling upon the tools of scientific hypothesis testing.

Insight, Revelation, and the Unknowable

Ask any scientist what lies at the core of her work, you will learn that it is not the experimental test of the hypothesis — although that is where most of the time and money in science go. It is the idea, the mechanism, the insight that justifies all the rest of the work of science. The moment of insight that reveals the new idea, where an instant before there was just fog, is the moment when the unknown first retreats before the creativity of the scientist. Here, then, is the first door into the unknowable: where does scientific insight come from? Surely from someplace currently unknown. Let us consider the possibility that scientific insight, like religious revelation, comes from an intrinsically unknowable place.

It is a safe bet that working scientists would agree to the notion that there is a lot we don't know yet, and that the boundary between the known and unknown that science pushes back is the shoreline of a small island floating in a vast sea of the unknown. Let us say — make the further hypotheses — that the sea of the unknown is not the edge of everything, and that the unknown itself is wholly bounded, blurring into an intrinsically inaccessible and immeasurable unknowability. Then science would still be increasing the territory available to the world of the understood. As the length and complexity of the shoreline with the unknown grew in step with every discovery, there would still be no edge to the unknown except the unknowable. The enterprise of science would be assured of a limitless future of successes, none of them ever bringing the unknowable any closer.

Can these hypotheses — that the unknowable exists, and that it will remain unknowable — be tested through the methods of science? Probably not, as they posit notions that resist testability. But they are nevertheless a fair representation of worldwide human experience outside science. It will be my first task to make the case that they are, as well, consistent with the actual experience of scientists, if not the institutional ideology of organized science. I hope then to demonstrate that, at least for the life sciences I am most familiar with, there is a way to practice the enterprise while also acknowledging that the shoreline may be remodeled but that in the end the sea is not drainable.

I can anticipate the response of many to what I have said so far: to beg the question. The unknowable is not a category that gives itself easily to demonstration of its existence. If it were a mental quirk only, a fantasy not worth worrying about, an idea of something that cannot be, then that would be a sufficient answer: No unknowable, no problem. The problem with that glib answer is that science itself depends on the periodic emergence of the unknowable for its own progress.

There is no way to think through a good idea in advance; insight is not a phenomenon subject to prior scientific analysis. At every instant of insight, every moment of Aha! what had not been conceivable becomes clear. Where was the idea before it was thought? Only afterward, once it was thought, can science begin the determination of the known from the unknown, using the idea as a guide. But before it was thought there were no tasks, there was no path, no idea that a question even existed to be asked.

The unknowable is worth a scientist's attention if for no other reason than that it is the source of insight, the irrational part of science that has no chance of being brought under rational control. Moments of insight in science are not reproducible, neither is their occurrence modeled by any hypothesis of its own. As scientific insight cannot be harnessed to the engine of experimental testing, each occurrence may as well be a gift from an unknowable source. Good ideas emerge in the mind of a scientist as gifts of the unknowable. They are not, as data are, simply trophies of a struggle with the unknown.

Insight Is No More Reproducible Than Revelation

The essence of the measurable is reproducibility; insight is by definition not a reproducible thing. Recall how few such ideas have come to any of us in the hundreds of years we have been trying to understand the world and ourselves through science. Yet without moments of insight that emerge from nowhere, science bogs down in mindless repetition of acts that look serious but cannot be in the service of anything except confirmation of what is already known.

Scientific insight is not the only example of such a gift from the unknowable. Other events — also occurring rarely, inexplicably, unpredictably — can give meaning to our lives, just as scientific insights can explain the world outside ourselves. By meaning, in this context, I mean a new understanding drawn from the internal emotional content of the experience, not the intellectual understanding that may follow as it does when experimentation proves a scientific insight to be useful. Meaning, purpose, teleology, the end of things: these are not notions that we naturally associate with science. Such experiences are commonly called religious.

Yet the central event in science — the sudden insight through which we see clearly to a corner of what had been unknown — is so similar to these religious experiences that I see only a semantic difference between scientific insight and what is called, in religious terms, revelation. That difference remains small, whether one says that insight or revelation both come from nowhere interesting, or that they come from the unknowable that surrounds all that can be known, or that they come from God.

The differences between science and religion that have crystallized and reified into a wall separating the two do not lie in the semantic difference between insight and revelation. Whether prepared for or not, prophetic experiences and scientific insights will occur with similar rarity, irrationality, and unpredictability. The real differences grow from the different uses made of scientific and revelatory insight. In both insight takes the form of a vision of an invisible and hidden mechanism. In science such insights are made into guides for learning how nature works, thereby reducing our ignorance of the world around us. Guiding the formation of religious obligation, revelatory insights are prerequisite to the rituals and observances of a religion, which ease the burden of living by lifting a felt ignorance of the purpose and meaning of our mortal lives.

In all organized religions I am aware of, revelation takes the form of a sense of being overwhelmed by sheer feeling arising within without reason nor cause. Just as a scientist prepares for insight by deep immersion in the study of what has been dragged out of the unknown by her predecessors, a person adept at religious insight — a holy person, a prophetic person — may prepare by study of earlier revelation and prophecy, and by trying to be alert to the moral or lesson taught through what might be — to an unfeeling observer — just a coincidence.

Although both science and religion presume that the territory of the unknown is vast, most religions are far more comfortable with the notion of a residue of unknowability than are most sciences. Many practicing scientists instead believe — they would say they know — that what is not known today must and will be known tomorrow, or the next day, and that this will go on until everything is known.

The notion that nothing exists except what is knowable is wholly unprovable. Holding on to this belief in the absence of any way to test it through experimentation, and despite the counterevidence of scientific insight itself, puts science at the risk of being trapped in dogma. Like the worst of religious dogmas, the insistence that everything is knowable is an unprovable position assumed in the face of the evidence of the natural world. In this case the evidence includes the fact of un-controllable insight as the wellspring of scientific discovery.

Scientists will argue that the reproducibility of scientific ex-periments assures that science as an enterprise can always be brought to internal consistency, while religions, free to call upon individual revelation and unreproducible miraculous events, necessarily fall into contradiction with one another and thereby cancel any reason for a sensible person to take any of them as seriously. In a negative template of this position many people of faith will argue that science is a fragmented enter-prise unable to paint a coherent picture of the natural world, limited by conflicting and inconsistent models and the finite limits of a mortal mind.

Whereas many scientists cannot really accept that anyone could believe in a way around mortality, and though many re-ligious persons cannot really believe that any serious person could fail to experience those feelings, some people — I am one of them — choose to carry both sets of thoughts at once. From the point of view of a scientist who is also a religious person — or of anyone else willing to allow the irrational portion of his life to be admitted to the discourse — religions respond to a small number of universally felt human experi-ences, the most easily recognized across all cultural boundaries being the obsessive need to somehow come to terms with the rational vacuity of one's own mortality and the recurring need to vest one's life with a meaning that transcends it.

Accepting the Irrational in Science and Religion

In my book *The Missing Moment* I concluded that current sci-entific studies of the brain and the mind require us to ac-knowledge that science has an irrational component, and that scientists are likely to experience this irrationality as the same waves of awe, joy, fear, or wonder that can overtake a religious person or even the "oceanic experience" of a shared, external, unknowable presence that Freud protested too much he could not feel.

The barrier erected by scientists who push aside, deny, or ignore these irrational states of mind is an artificial unnecessary one, built on denial of the reality that their own work depends upon uncontrollable and unpredictable moments of insight. The same artificial barrier is maintained from the other side with equal futility each time the resultant discoveries of science are denied, ignored, or pushed aside by people anxious to protect the same irrational states of mind so precious to them in their religious faith.

To dismantle the wall from both sides, both camps will have to admit what they must already know: the reality of irrational inward experience. They will have to acknowledge it as the source of the unexpected and unpredictable insight upon which both organized science and organized religion depend. Such admissions will not come easily. Characters like me are not at all used to putting religious feelings in the foreground and, rather, have the habit of pushing our feelings away, repressing them into unconscious reservoirs from which they may emerge but where they do not interfere with the dream of lucid rationality.

This makes speaking about religious feelings in an academic setting particularly tricky. Scientists and others who use their powers of repression to avoid accepting the reality of religious feeling, or even its origin the natural world, tend to have great difficulty accounting for such feelings even in themselves. Not just moments of insight and revelation but other feelings as well — emotional states that overtake one, unbidden and unplanned by conscious rational anticipation — seem to be a different order of phenomena than those easily studied under reproducible conditions. It is extremely difficult to do a controlled experiment on feelings.

In terms of the expected behavior of scientists, strong feelings as such are also in bad taste. Data have to be examined in terms of the model they test, and models as well as data have to be able to stand on their own in the eyes of other scientists. This situation too has its mirror image in organized religion, where a spontaneous feeling of disbelief or doubt in the face of incomprehensible evil or simple bad luck may not be easily squared with the presumption that we are moral beings in a moral universe. Neither can all the unwanted strong feelings associated with love, aggression, or, of course, death be fitted into most religious frameworks of expected right conduct. Too much doubt is as much in bad taste from a religious person as too much enthusiasm from an overeager experimenter.

The Avoidable Risk of Dogma

While insight and prophecy may both visit a single person, neither organized science nor organized religion expect each of their members to share in the prophetic experience. Rather, each transmits the gifts of its most insightful leaders. As those of us in a university know better than most, teaching is a human interaction rich in emotion, and therefore subject to abuse. The abuse of the teaching function takes the same form in both organized science and organized religion: what begins as the fully engaged experience of meaning can be compressed, through unfeeling teachers, into flat cold dogma. All religions have their dogmas, and so do all the sciences. Dogma always takes the same form: do not ask, just memorize; do not feel, just do.

Perhaps the most cogent reason to seek a reengagement of science and religion is that each can help the other to shed dogma that constrains their adherents, freeing each scientist and each person of faith so that they may be available for their own rare but precious moments of revelatory, insightful understanding. If that seems too ambitious an enterprise, it is because one has forgotten the great harm that can come from unchecked dogma coupled with even a slight excess of aggression. The usual failure of dogmatism is self-inflicted ignorance, but there is a pathological variant that insists on teaching ignorance to others, or making others pay for one's own willful denial of an aspect of the world.

As Thomas Cahill writes in the introduction to his problematic book on Judaism from a Christian point of view, *The Gifts of the Jews,*

The teacher . . . is . . . someone who attempts to re-create the subject in the student's mind, and his strategy in doing this is first of all to get the student to recognize what he potentially knows, which includes breaking up the powers of repression in his mind that keep him from knowing what he knows. . . . Why are belief and disbelief, as ordinarily understood, so often and so intensely anxious and insecure? The immediate answer is that they are closely connected with the powers of repression I referred to earlier as being the teacher's first point of attack.

Dogma is unlikely to be a part of any serious science, nor of any serious religion. As dogma is the enemy of insight, it is as well the enemy of both scientific progress and religious revelation. The front page of the *New York Times* of August 12, 1999, provided a clear example of the consequences of pathological dogmatism. Two articles that day seemed unconnected except by coincidence, but together they taught an important lesson. The lead on the right side read "Board for Kansas

Deletes Evolution from Curriculum: A Creationist Victory." The lead on the left read "Man with a Past of Racial Hate Surrenders in Day Camp Attack." I doubt that this man had any notion of evolution — for or against — when he "issued his message to America by killing Jews," nor do I imagine the Kansas Board of Education had any differences between Jewish and Christian children in mind when it removed from them all any reward for learning about evolution and the central value that natural selection places on the uniqueness of each person and the subsequent diversity of our species. Nevertheless, both articles were about just those dogmatic matters. In both cases dogma had had the power to deny aspects of our common ancestry and, in so doing, to override the revelatory insights of both science and religion.

Natural Selection Without Dogma

Free of the dogmas of both science and religion, any curious and self-aware person should also be able to know clearly the facts of nature through science and to feel clearly the meanings and purposes of those facts through religion. But, for me, this has not been so easy. I find it particularly difficult, for instance, to connect my place in the universe and my reason for being here with certain facts uncovered by my scientific colleagues. The molecular biology of evolution, in particular, has uncovered facts about me and the rest of us through the experimental testing of scientific insights that fit badly, if at all, into my religion's revelation of meaning.

Since Darwin wrote *On the Origin of Species* 150 years ago it has been a strong fact of my own science that the tens of millions of kinds of living things that seem to breed true over time — we call them species — share a common ancestry.

Out of this fact of common descent comes the unlikely notion of a species having a finite lifetime. Species are not the stable entities they seem to be; each collection of slightly different individuals capable of producing fertile offspring, despite their differences, is transient. As the variants in each species jostle with each other for food, space, and a fighting chance for their offspring, some will survive and others not; in time species will change as a result. Eventually — a species' lifetime that can run for many millions of years or can be much shorter — it will either die off as it is pushed aside by other species invading its ecological territory or be supplanted in that territory by a new species emerging from a minority of its members.

The two notions of common descent and species mortality were well laid out by Darwin and confirmed by others immediately thereafter, but it took one more century for another unintuitive insight to complete today's scientific understanding of the origin of species: the fact that inherited changes in a chemical called DNA could accomplish much of what the Darwinian ideas of common descent and origin of species required.

DNA, assembled in long, informationally rich threads called chromosomes, forms the genome of a species. A genome carries all the information necessary for the construction of each organism in a species; organisms in a species vary from one to the next because their genomes vary. Each of us carries our version of the human genome in each of our million billion cells. Each of us — and each individual in every species — becomes slightly different from the others because the copying of a DNA genome from generation to generation is never error free.

When error generates new sequences of DNA that happen

to encode enhanced survivability in a species' offspring, a new fertile population may emerge from an existing species. In time it may become a new species, replacing its parent; species themselves are thus no different from the individuals that make them up: like individuals, species are born, live, and die. That is why either replacement or simple disappearance is the certain fate of all species, including our own. These facts from science tell us, in other words, that our species — with all our appreciation of ourselves as unique individuals — is not the creation of design but the result of accumulated errors.

The Imperfectability of the Living World

The scientific confidence in these facts about our own origins and our own eventual fate is buttressed by other, equally well-documented facts about DNA-based life on Earth. In earlier times there were no humans, and in even earlier times there were no mammals, nor vertebrates, nor any organism bigger than a single cell. From those earliest times until now, all that we might want to think of as progress has been simply the selection of one subset of DNA sequences or another from a constantly refreshing pool of copying errors. We can be fairly certain that replacement or death will be the fate of all humanity as a species, just as death is the certain fate of every person.

The methods and strategies of science have thus brilliantly succeeded in explaining how we got here and where we are going next, and the explanation seems to leave absolutely no room for meaning or purpose. A mutation just happens to land in the sperm or egg that will make one individual and not another; no design to its occurrence is either necessary nor even demonstrable. This most successfully defended null hypothesis of science has been so amply confirmed that there is no longer any reason to doubt it.

The living world, ourselves included, is intrinsically imperfect and intrinsically unperfectable. It changes, but even the changes that make each of us individually unique and interesting to each other are meaningless differences in DNA, creating the differences among us toward no purpose beyond the possible improvement in survival of one or another particular version of DNA over time. Even that imputation of purpose to the data may be unjustifiably teleological.

I am not exaggerating the seriousness of this problem: scientific insight into the meaninglessness of DNA-based life is not simply missing meaning. It is the demonstration that a satisfactory, even elegant explanation of the workings of this aspect of nature actually conflicts with the assumption of purpose and meaning. There is neither the need, nor any sign, of an unknowable designer in these data, nor any sign that greater meaning and purpose will one day be drawn from these data.

Honest scientists know their limits. Newton excused himself from the task of finding meaning in his discovery of the laws that govern the movement of stars and planets by saying, "I have not been able to deduce from phenomena the reason of these properties and I do not feign hypotheses." Unless we force science to do just what Newton did not deign to do and simply articulate our wishes as if they were in the data, though they are not, we must accept the meaninglessness and purposelessness of our presence on Earth as the verdict of testable science.

Yet you may, as I do, find it impossible to understand your place in the universe on these facts alone, and find yourselves asserting with me the irrational certainty that there must be meaning and purpose to one's life despite these data. With those assertions we can begin to take down that wall, by asserting as well that the irrational certainty meaning exists — based not on data but on emotional necessity — is itself a data point about the living world that can and must also be understood. Many scientists do not agree with any of these assertions, a personal choice they are certainly entitled to. Such a data-free choice, as irrational as any of mine, is not a problem so long as it is not illegitimately given the weight of a further fact of science. As is any act of free will, their choice is no different in its emotional inward origins than my own, except that theirs does not allow for any further discussion.

Denying the Unknowable

Steven J. Gould of Harvard University and *Natural History* magazine is a lucid writer and a very serious student of evolution, so it was all the more remarkable to see him choose to refuse to accept the parity of unbidden religious revelation and unbidden scientific insight in a surprisingly conflicted 1999 editorial essay for *Science* magazine. Taking on the task of helping scientists to make the facts of evolution palatable to religious people, Gould seemed at first to agree that the acquisition of meaning and purpose through inward and emotional religious experience is real even for scientists: "Factual nature cannot in principle answer the deep questions about ethics and meaning that all people of substance must resolve for themselves. When we stop demanding more than nature can logically provide . . . we liberate ourselves to look within."

Clearly, then, he has agreed that questions of meaning — like my question of how to make sense of my own life in a world brought about by natural selection — can be answered only by "looking within," and that answers cannot be had for questions like these from any corner of science. But Gould does not follow the concessions of his heart. Instead he reverses course and ends his essay with a paean to the facts of evolution. He accepts our evolved brains and bodies as real, but rejects some of the equally real feelings generated within these evolved brains and bodies — such as the certainty of a transcendent purpose in the face of all evidence against it — as meaningless, parochial, wordplay: "Let us praise [evolution] — a far more stately mansion for the human soul than any pretty or parochial comfort ever conjured by our swollen neurology to obscure the source of our physical being, or to deny the natural substrate for our separate and complementary spiritual quest."

For a religious person, to "look within" is to look to one's feelings for the knowable signs of intention and meaning, regardless of the unknowability of their source. Such a conviction is not properly addressed as a "pretty or parochial comfort." The notion that the Unknowable must not exist if evolution does is not pretty, but it is parochial insofar as it lets the facts of evolution set the terms of what we can feel as well as what we can know.

The parochialism of science is to erect this barrier against thoughts and feelings that are not rational, so that what we cannot rationally understand is not allowed to exist. Perhaps the real "trick of swollen neurology" is Gould's trick of raising this barrier against a significant aspect of that neurology's output, even as he appears to be completely open-minded.

Gould, a fastidious scientist and the preeminent popularizer of the evidence for natural selection and evolution, had staked out a simpler position earlier, that science and religion have little or nothing to say to each other. He argued then that the sciences, which deal with what can be known through direct experience of the world, and the religions, which deal with what can be known by direct experience of inner feelings, are so completely separate as to be distinct and independent magisteria, with no point of contact.

That seems to me less heated but no more fair. His recent editorial in *Science* suggests that Gould has perhaps come to agree that the surface of contact between science and religion touches not only the common experience of revelatory insight but also a larger set of shared feelings. These feelings too are data points consistent with the notion that the meaning and purpose of the natural world can — some would say must — extend beyond a description of its workings.

Accepting Meaningless: Religion as a Parasitic Meme

Richard Dawkins of Oxford is the inheritor of Huxley's mantle — pulpit — as the current expert on natural selection who best articulates a vision of science that would abolish all religious insight and reduce all irrational certainty of purpose and meaning to meaninglessness. A quarter-century ago his landmark book *The Selfish Gene* set out the case for a science — and a world — free of all unknowability, and so also free of all meaning beyond its own orderliness, whether from religious impulse or from any other spontaneous modulator of free will. It took him only three steps. Whether you see it as a tour de force or catastrophe, it is worth attending to his line of argument.

First, he lays out the great and polished set of data confirming the basic premises of Darwinian natural selection: that no ideal form exists for any species, that each individual of every species differs slightly from all the rest, that whenever any of these differences leads to even the smallest increment of possibility that its progeny will survive and be fertile in turn, the variant's DNA will prosper at the expense of other versions of the species' genome, and that therefore all living things including ourselves are simply survival machines.

Next he argues that we too must all be Dawkins survival machines, thinking we have free will but actually — no matter what we may think or feel — devoid of any other purpose but to survive, gifted with no other special attributes but a set of variant DNA sequences compatible with differential survival, and as certain as any other species eventually to be overtaken and supplanted. Among these DNA sequences must be ones that give us the ability to teach and learn through the full use of language; these have undoubtedly conferred on our species a major portion of the additional survival capacity that has helped us to surge upward in numbers by a thousandfold in only the last ten thousand years.

Dawkins accepts that our brains are indeed special, but argues that what makes them special is not the free will to choose from among personal insights and revelations, but only the capacity to learn and propagate the ideas of others. Then he drops his bomb: he makes the hypothesis that our species' unique, idea-driven survival strategy of teaching and learning means that our societies are no more than the test-beds for Darwinian natural selection in the mental world of competing variant ideas.

Ideas that survive the filter of natural selection in our societies to be taught, learned, and remembered by large numbers of people he calls memes, by analogy to DNA sequences that survive and are called genes. Successful memes will, in this model, have survived for the same reason, and toward the same goal and purpose, as successful genes, to wit, no purpose at all beyond their own survival.

The notion of a meme is a very successful meme, as witness my use of it in this book. So, to Dawkins, are all the ideas of meaning and purpose attendant on any and all religions: religions are collections of successful memes. And as all successful memes are the products of a novel version of natural selection taking place only in societies of people, no ideas need have any greater content or meaning than that which assures their own survival. In other words, our most precious ideas need be no more than parasitical memes, infecting our minds and behaviors the way DNA-based parasites infect our bodies.

Parasitical memes do resemble viruses quite well. They are the ideas that command their own propagation, preserving themselves in individual memory and making themselves be taught and learned, in turn, even at cost to the person who serves as their carrier and vector. In the world of ideas and their propagation by teaching and learning — the world of memes — a successful parasitic meme is precisely what a virus is in the world of DNA and its propagation by replication. Real viruses are parasites because they use their DNA genomes to force cells of the body to switch their work from the body's needs to the virus's. The genome of the virus encodes information — a string of DNA — that can command the cell in which it lands to copy itself, and to read from these copies the further commands to produce the virus's coat, and then perhaps even to kill itself so that the virus can be released to infect other cells in turn.

Like the variant DNA sequences of viruses that sustain themselves because they enable their progeny to survive despite the body's defenses, parasitic memes are products of selection for the best chance of slipping though the mental equivalent of the immune system — the sense of what is worth remembering and doing — to be remembered and taught. A chain letter is an arresting but trivial case for Dawkins's all-encompassing view of memes. In a paper in *Nature* a few years ago, Dawkins and Oliver Goodenough described one such mind virus, the "St. Jude chain letter."

Chain letters promise happiness to the recipient on condition they are sent out to a multiple of people in turn. The St. Jude letter promises "good luck in four days" for those who send on ten copies to their friends and relatives. No strings here, no tricks, just the antidepressive effect of irrationally believing you might control your luck by helping others to control theirs. In their note the authors point out that though they turned out to be immune to the virus, they did experience "waves of mild, irrational anxiety on deciding not to comply." Indeed, they managed to comply with the virus after all, by having it reprinted in *Nature*, albeit in a format that might be thought of as a vaccination against it.

Just as the deepest problem with natural selection is that it reduces all meaning in life to a set of ways for DNA to experiment with making more DNA, the deep problem raised by the notion of parasitical memes is that they too need have no meaning beyond their own propagation. In a world of meaning there is nothing unexpected about a person being taken with the meaning of a set of ideas and remembering or teaching them. But, in a world of memes, even those ideas need have no further meaning than their own propagation. That is creepy indeed: perhaps the best way to see the problem is to imagine that Dawkins himself was infected by the meme called "The Meme" and that, even as I write these chapters, I am infecting you with it, to no other end but its own survival!

Dawkins seems to exempt science from this reduction to meaninglessness: the memes of science would be surviving by helping us to unpack aspects of the orderliness of nature, whereas all other sets of teachings that give meaning and purpose to the world beyond its orderliness would be reduced to the status of a cluster of parasitic ideas, meme viruses, so to speak. But the world of memes absorbs even the scientific enterprise with little effort.

In the world of memes the scientific method — the making and testing of hypotheses about the natural world — could not be the source of meaning or purpose, because all its insights would be no more than examples of success in the mindless, meaningless competition for survival of one idea over another. And that would not be a problem, because in that world all there is to any idea at all is survival. Moreover, what is the testing of hypotheses, if not a competition among memes for survival in human societies? If the world of ideas is no more than a world of memes, then, like any other human enterprise, science would be quite completely and permanently closed to the possibility of meaning or purpose.

Free Will: Choosing an Irrational Path to Meaning

At this point I must either begin to get us out of the world of memes or close up shop. There are two steps that lead back to the world some of us prefer to live in, a world bounded by an unknowable source of real meaning and purpose. The first — symbiosis — is for those who are comfortable with meaning but uncomfortable with an unknowable source for it. The second — free will — requires at least a tolerance for the irrational.

It is quite true that all parasites may be fairly described as successful outcomes to a process of mindless selection for survival, but survival may also depend on less selfish strategies as well. Some simple cells, for instance, are not parasites but symbionts. Like parasites, they are propagated with the help of a host and they would not live long except in the body of the host, but they in turn provide their host with something essential for its own survival. The bacteria that ferment the hay in the rumen of a cow are symbionts, and so are the ones that protect our intestines from true parasites. It may be hard to accept that bacterial symbiosis is part of our good health, but if you ever have the bacterial flora of your gut displaced by a parasite, you'll have firsthand knowledge of the difference between the symbiotic and parasitic strategies of survival.

Memes can be symbiotic as well as parasitic, surviving by sticking in our memory and emerging through our teaching while also giving us something we want and need in turn. Within the world of memes a mental symbiont might have survived for any of three reasons. Two are uninteresting: a meme might be advantageous to the species and have spread because individuals thinking it do well, or it might have been a harmless piggyback to a parasitic meme. One additional possible reason for a meme's survival is interesting: it might benefit each of its hosts, but in a way directly in opposition to what might help the species as a whole to survive; in other words, some memes may be antiselective in classic terms as well as individually symbiotic. The teaching and learning of instants charged with the presence of meaning beyond understanding, meaning that commands us to change our behaviors in ways that do not serve the dictates of natural selection, would qualify as examples of such antiselective symbiotic memes.

The first two sorts of memes allow the purposelessness of natural selection to explain what would otherwise be meaningful altruistic impulses and religious rituals. But the third, unnatural, meme survives in us despite, rather than because of, natural selection. Why do such memes survive? Who knows? Lacking data to the contrary, I would argue that antiselective symbiotic memes survive because people have the free will to choose to remember, teach, and learn them. Free will is the irrational key here: the fact that we find within ourselves the capacity to choose — on any grounds at all, but especially on irrational grounds, against judgment, against data, against survival, against reason, even against death — to learn, remember, teach, and observe one meme and not another, returns meaningfulness to us.

Among our choices may well be successful unnatural memes of true religious meaning, but that is not the issue. The irrationality of free will is a fact of the natural world. This irrational aspect of the way things are is, like insight and revelation, yet another emergent, knowable aspect of what is otherwise unknowable.

Judaism and Free Will

Judaism places a high value on the notion of free will. The entire framework of Jewish understanding of our place in the world, our responsibilities to God, and to each other, is built upon the unique human capacity to make irrational choices as well as calculated decisions. Decisions may be made by many species, and the selective advantage of brain wired for logic is plain, but only a person can make a choice despite calculation rather than because of it.

In the Jewish tradition the God who has existed since before time and the universe began created both time and the universe in order to have, in time, creatures — the word means things created — with free will who could then choose to say thanks for their and the world's existence. For thanks to be proper and meaningful — the proper form of thanks is to bless God — these creatures would need absolute free will to choose whether or not to do so.

Hence the unavoidability of randomness, accidents, and, for

that matter evil, in religious terms: all must be allowed to result, whether by the wrong human choice or by truly random
occurrence, because to allow any to be preventable by prede-
termining human choice would be to gut the purpose of the
creation. The absolute requirement of human free will in this
religious vision shifts human choice into the foreground and
mechanisms of natural selection that yield a person who can
make the unexpected choice into the background. This set of
unprovable assumptions — so bizarre in their distance from
anything reproducibly known through science and yet so familiar in their high regard for the critical step of insight in science — validates meaning and purpose in a living world that
is the product of the random processes of natural selection.

This line of argument is articulated beautifully in Adin
Steinsaltz's book *The Strife of the Spirit*, in the essay "Fate,
Destiny, and Free Will." I had not yet read his essay when he
and I first talked about these matters. Having just read an earlier article by Richard Dawkins, I was quite astounded by his
capacity to reduce religious thought to an especially successful
kind of ideational parasite. Rabbi Steinsaltz's answer was to
give me a reference to his essay, with the passing remark:
"God says, 'Get me a thinking creature, I don't care how.'"

From that religious perspective Dawkins can be wholly
right about memes but also wholly without force when it
comes to the question of purpose: natural selection would then
be a wholly random-driven mechanism that would eventually
yield creatures with the capacity to propagate memes, but also
the free will to choose among them.

In specifically Jewish terms, then, it is the God-given inexplicable reality of free will that allows us to accept a meme —
or a revelation — or not. That choice — available not just to
Jews but to all people as their birthright — makes us the active
determiners of our fate rather than the passive vectors of our

memes. I would like to conclude as well that free will is also a fact of nature that would absorb the notion of memes, placing them — and all other experimentally derived facts of nature — on a par with the unknowably derived felt impulses of revelation. That would leave Dawkins with all the credit he deserves, without leaving the rest of us bereft of meaning.

But what if Dawkins is right all the way down; what if the notion of free will is also just a meme? Then the decision to live as if free will were real — the flip side of Pascal's famous wager that it is a good bet to live as if Hell were real — would still give meaning to our lives. Twenty-five years after the publication of *The Selfish Gene* no religion seems shaken by the emergence of the meme of *memes*. This triumph of creative irrationality over the unbearable tells me that, just as Dawkins cannot say from whence his idea of memes emerged, all of us — those commanded to choose and those choosing alone — may still choose to ignore it in order to have meaningful lives.

Medicine's unnatural quality — which it shares with other sets of symbiotic memes that work against natural selection, like certain religious rituals — is its insistence on the essential importance of a single person's pain and suffering. Pain, suffering, unreasonable maldistribution of good and bad fate: these are the very stuff of natural selection, the visible expression of the random genetic variation that provides nature with the eerie capacity to produce some living thing that will survive any contingency. To work against these aspects of life is to work against its deepest mechanisms, and also to work against the meaningless of these mechanisms.

We have returned at last to Tolstoy's questions of what shall we do and how shall we do it. In closing, here are my answers, in the medical context of the next two chapters.

First, things are the way they are for no knowable reason.

Second, because of this, we may by the constant exertion of free will choose to know and to act according to both our intellectual powers of understanding and our emotional powers of feeling.

Third, in this way meaning and purpose may be vested in our lives and in our actions, without the penalties of denial and rejection.

Fourth, nothing intrinsic to either the scientific method or the discoveries of science precludes any person from holding to any religious faith.

Fifth, religions that have emerged from different revelations may engage the world of data in wholly different and novel ways.

Sixth, because much of today's science as we practice it here has its roots in the work of people who were deeply immersed in the three faiths that find their origins in the story of Abraham and his descendants, as recorded by Moses, the deep separation of science from these three religions in particular should be easily reversible; that is, what the Enlightenment tore apart, enlightened people can freely choose to bind together again.

In the next two chapters I will follow out a few implications of the tragic reality that fertilizes the common ground of medicine and Judaism: that, in the only world we can know, death must prevail. The fact of mortality — so hard to understand even though it is accepted by my religion and demonstrated by my science — brings the worlds of "How?" and "Why?" together. When "How?" and "Why?" are kept apart — as they are often by the shared agendas of basic science and genetic medicine — the risk is great that medicine will be used not to help us to acknowledge death but to deny it.

chapter two

**The Meaning Is in the Order:
DNA-Based Medicine**

> To believe it is possible to know a universally valid truth is
> in no way to encourage intolerance; on the contrary, it is the
> essential condition for sincere and authentic dialog between
> persons.
>
> — Pope John Paul II, *Encyclical Letter Fides et Ratio*

I N THE REMAINDER of the book — orig-
inally presented as the second and third
Schoff Lectures — I make the case that the
current practices of my science of molecu-
lar biology of disease and of my religion of Judaism would
contribute to an improvement in medical care if each were to
appreciate the insights gained by the other. This is so because
even though at first glance the working habits of the inter-
preter of Jewish written and oral law — the rabbi studying
Talmud — and the scientist whose work is intended to ameliorate or prevent disease — the doctor — seem to exclude an
overlap of interests, on closer inspection they have a lot in
common.

The Rabbi and the Doctor

The differences are clear: the rabbi studies a sacred book —
an accumulation of revelation and interpretation recorded

orally and then in writing over three millennia — to understand how to meet God's wish that His People choose life by interpreting how to live according to this book. He will spend his life trying to understand how to translate these words into actions consistent with an unknowable God's intentions, knowing his human abilities are far too limited to allow him the certainty that comes with understanding, but that his feelings are not too limited to allow him the certainty that comes with faith.

The medical scientist studies the living world, in particular the fraction of it available at hand, and finds in its regularities confirmation of laws that apply to the entire known universe as well as hidden mechanisms of astonishing beauty and wit. Using carefully chosen samples from the living world to test her emergent models, she continually deepens her understanding of how life works. For her the purpose of any aspect of nature lies wholly in its details; intentionality and meaning are never a part of any good scientific model.

Despite these apparent differences, the rabbi and the medical scientist share a great deal in their different ways of seeing the natural world. First, they agree that the living world can best be understood by humankind through the study of a text. In Judaism the text from which to learn why is called Torah. In molecular biology the text from which to learn how is a chemical one, made of DNA, writ small but at great length in our chromosomes.

Second, they agree that the text preceded the present time not merely by some large number of generations of humankind but by a gap in time long even in comparison with the lifetime of our species. In Judaism both the Unknowable God and Torah exist outside of time's passage, eternally present before the creation of the physical universe. In molecular biology the textual genetic information contained in nucleic acids — information we carry today in each of our cells — began as a set of molecules arising out of primordial muck around five billion years ago, soon after the earth — itself appearing from a point source of all creation some ten billion years earlier — had cooled sufficiently to carry liquid water.

Third, they agree that their current understanding of the text of their choice is woefully incomplete. In Judaism the age of direct prophecy has passed, leaving all Jews with a task of interpretation that shows no sign of resolution. In molecular biology the text within ourselves — the human genome — has not yet even been wholly carried from the chromosome to the computer. It is the focus and fascination of the day, and investments in decoding it — merely a prerequisite to the beginning of understanding it — cost billions of dollars each year. There are already signs of its being understood in ways that might permanently diminish someone's free will, but no signs of it being well enough understood to vest it with purpose.

Fourth, they share two beliefs founded entirely on faith, with no possible data to support either: that one day the text of their choice will be completely understood and that on that day death will have no power over us. Judaism looks to the day when the entire world is One, united in a complete understanding and acceptance of the unknowable, unprovable wish that humankind live according to Torah; on that day time will stop, history will end, a messianic era will begin, and death will have no dominion.

Molecular biology, no less as an act of faith, looks to the day when the entire human genome is captured in a computer bank, when each version of it in each of us, and each version of every gene in one of us somewhere on the planet, will all be logged and fully understood. On that day the workings of the body and the mind will be fully understood, the genetic component of anything human will be reproducible at will, and, again, death will have no dominion.

Regardless whether one, or both, or neither of these deeply held faiths is acceptable, avoidable, laughable, or necessary to any one person, my purpose for the remainder of the book is to build from the observation that common assumptions and common acts of faith make medical biology and one particular form of Judaism more closely allied than they seem at first sight to be.

Meaning, Medicine, and Transgenes

How, then, might the insights of Judaism and the discoveries of DNA-based medicine together give meaning to the otherwise purposeless sequences of DNA? Many DNA sequences might be made to serve our purposes, but for dramatic effect we will begin by considering the green-eyed frog.

If you go to the right basic medical science laboratory and ask the right people, you can get to see a family of frogs whose eyes look back at you glowing a fluorescent green. These frogs have green eyes because they have green nervous systems. Cut one open and its brain, with all the nerves that run to and from it, will glow with the same bright fluorescent green that shines out from their retinas through their eyes. Where do these frogs come from? What do they mean?

They are transgenic frogs, the product of a line of experimentation designed to illuminate — in the classic sense of the word — the assembly of the nervous system during the development of an embryo. Their purpose is to add to our store of knowledge about the underlying mechanisms that allow the descendants of a single cell — the fertilized egg — to each become different from the others, so that a developing embryo gains form and does not merely grow into an ever larger boring ball of cells.

No laboratory makes a frog of course; only the sperm and egg of two other frogs can do that. All animals begin when a single cell — classically, a fertilized egg cell — divides many times, each of its many descendants becoming different from the others. Laboratories can make clones of a frog by moving the nucleus of a frog's cell into the cytoplasm of an unfertilized frog egg and letting the resultant reconstituted cell develop into an embryo. But even a cloned frog grows from a nucleus that came from an earlier fusion of sperm and egg, and even a cloned frog that begins in a laboratory dish grows by the same process of differentiation among early embryo cells.

The differentiation of many different kinds of cells is paradoxical on its face because — just as every copy of a theater program is the same — every embryonic cell descended from a fertilized egg cell carries the same version of the genome as every other cell. How are identical sets of genes read so differently in the different cells of the embryo? By opening up the genes in the nucleus of a cell, scientists have elegantly resolved the paradox of differentiation and embryonic development from a common genome.

The DNA genomes of sperm and egg cells are sealed up, silent and unread, even though the egg cell's cytoplasm is packed with chemicals that can activate many different sets. Fertilization loosens the wrapping from the sperm and egg genomes and from the copies of DNA that will be made as the fertilized egg cell divides. The first few cell divisions partition the egg cytoplasm, so that each early embryo cell gets a different mix of the egg's gene-activating molecules. These different mixtures launch each early embryo cell along a different pathway of differentiation, making each the ancestor of one or another differentiated embryonic tissue.

Once an early embryo cell gets its maternal cytoplasmic road map, it and its descendants will continue to activate changing sets of genes, extending and ramifying differentiation at every generation. While some early genes activated by egg signals make the materials we see as the end-product of differentiation — the pigments of blood cells and hair cells, the long fingers of nerve cells, and the like — others can activate or turn off new sets of genes in turn, opening cascades of gene regulation that resemble the flow diagrams of a computer programmer. Some turn on or off other genes in the cell that makes them, while others make products that cross the boundary from one cell to another to trigger swaths of differentiation elsewhere in the developing embryo.

At no time are the differentiating cells of an embryo free from such regulatory introspection and conversational contact, so the embryo's tissues continue to emerge and sharpen with time. All tissues continue this conversation for the lifetime of the animal; there is the possibility that death itself is the last differentiation of all tissues. The mature brain — so poor in its ability to restore itself from a wound by cell division but apparently still dependent on new cells for the tagging of memories — is also most adept at continued differentiation in response to the outside world's stimuli. We call these differentiations learning.

The technique of making a frog's developing nervous system visible is the culmination of many years of work that began with the technology of reading DNA of a fertilized egg or of a cloned nucleus as a text, of editing it, and of testing various edited versions for their ability to successfully carry out the agenda of embryonic differentiation. One such edited frog genome led to the discovery of a gene whose continued activity is required in the cells that become the nervous system. This gene was then spliced to the gene encoding a green fluorescent protein found in the sort of jellyfish that lights up the waters of New England beaches at certain times of year, the aptly named Green Fluorescent Protein, or GFP.

Frogs whose genomes carry the edited GFP-tagged version of the nerve-specific gene emit green light from each nerve cell as soon as this gene is activated. Cells that are or will become the frog's nervous system shine out, and no further manipulations are necessary to see them. From the earliest embryo to the mature frog, the entire developing nervous system shows itself in the living organism without the need for any further anatomical dissection or staining.

With the nervous system laid out to the observer like a city by night, a new sort of experiment can be done. Scientists have begun to rewrite the genome of the frog not only to see development in action but to modify its course. As they find more of the many genes required for normal neural development, they will be certain to use GFP tags to illuminate the consequences of inserting additional mutations that alter the nervous system in subtle or dramatic ways.

Transgenic People?

I tell you all this not because a green-eyed frog has religious significance, nor because it lends itself immediately to the task of discovering meaning in nature though medicine, but because nothing about the green-eyed frog is restricted to frogs. Mutant frogs, green eyes and all, may be unnatural, but each is also uncannily familiar, because people and frogs have a common ancestor. What can be done in an egg that will become a frog can be done in an egg that will become a person. We and frogs, and roses — and every other living thing — all are the bearers of variant DNA sequences descended through an un-broken history of DNA replication and DNA error from common ancestral species.

These insights of Darwin's — the ideas that all species might have a common origin and a shared ancestry — kept him from publishing his ideas for twenty years. His reticence was not modesty, nor was it motivated entirely out of fear that his ideas might somehow be repudiated despite his evidence. He faced the difficulty of accepting that his ideas applied to our own species: not the notion that we are the descendants of a now dead primate but rather the process of natural selection that selected the variant heritable traits making a person — and the other variants that make a chimp — did not have any detectable preference for us. It is natural selection's total dis-interest in us that strains its acceptability today, certainly among people of most religions but also among even some sci-entists, because it seems so contrary to common sense that hu-manity should not be in any way special.

Though life has been present on this planet for around four and a half-thousand million years, our species is very new, having first appeared no more than a few million years ago. About five hundred million years ago there were no frogs, no people, nor even any mammals. But there was, then, some sort of animal species from which emerged — by natural selec-tion — the ancestors of both frogs and people. Whereas that last common ancestor had its own genome, and as we and frogs are merely the products of somewhat different variant versions of that ancestral genome, the genes that are required for the assembly of a frog's nervous system are quite similar to the genes we use to assemble our own brains and organs of perception during our own nine months in utero. Mutations in the frog's nervous system may well mimic the inherited problems humans face, and understanding a frog's genes may well tell us something about the differences in the way different people perceive, think, and — most important for our purposes — feel.

Frogs are such a good stand-in for us, their genes are so close to our own, that even today nothing technological prevents using the technology that produced them to produce a fluorescent green-eyed person. Why not just do it? Human eggs are not hard to get; they are lying about in excess in laboratories that specialize in in vitro fertilization for couples — or, more precisely, women — who cannot for whatever reason conceive a baby in the ordinary way. The trip from in vitro to in utero is a bit more dodgy, but here, too, there is enough unfairness in the world that a rather small amount of money has often been sufficient to gain the use of a young woman's body for a nine-month term. We know which human gene to attach a GFP sequence to and how to put the edited version in place of the ordinary one in a human genome. As an experimental tool a GFP-fertilized egg of this sort could allow us watch the development of the human nervous system from its earliest moments after conception. Imagine the tour de force of having helped to make a baby whose eyes glowed in the dark!

I would be shocked if anyone were not shocked by now: clearly this is not going to happen in the normal course of scientific events. But, again, why not? First, because today such a manipulation of the human genome would be unethical. Ethics require no religion, that is, no meanings to be put in place by an unknowable God outside the data; they are a set of professionally sanctioned, agreed upon principles of right conduct. German physicians who carried out euthanasia in their hospitals from 1933 until the outbreak of war with Poland in 1939, and even those who carried out experiments in camps until 1945, could and did argue that they were working within the ethical norms of their profession.

Ethical norms can change with experience: since the Nuremberg Trials of 1945 focused the world on what had passed for medical research in the time of the Nazi regime in occupied Europe, the medical profession has agreed that it is unethical to make someone the subject of an experiment without that person's free consent or to continue an experiment on a person who wishes to be free from further participation. A fluorescent green-eyed child emerging from an editing of its egg's DNA would have every right to complain on both counts, as he or she would have no recourse but to live an entire life within the experiment's parameters.

But what if the gene to be edited were not just any gene active in nerve differentiation but one responsible for an otherwise avoidable inherited disease of the central nervous system — Alzheimer's Disease, say, or Huntington's chorea? And what if the editing were not to make nerve cells glow green but to insert a functional version of a damaged gene? Would it be outside the pale of medical ethics to accede to parents' wishes and to try this pathway to the birth of an unafflicted child, who would in this way inherit an otherwise wholly authentic genetic ancestry from its genetically afflicted parents? That might very well push the argument over and make the experiment medically ethical, if barely.

Whether or not to do such an experiment is not only a matter of rationally negotiated, agreed upon ethics. Isaiah Berlin's notion of the scientific fallacy — the presumption that what can be known through science is all that can exist — works against the legitimacy of any intrinsic moral core to medicine, as there are never going to be quantitative data in support of any guidance in distinguishing right from wrong. It is as much, or more, a matter of choice, an opportunity to transcend balanced consideration and respond instead to felt right or wrong. Religions come into play at such moments not to supplement the ethical rules of professional conduct with other sets of ethical rules, nor to impose a set of actions, but as a source of guidance, helping to assure that the choice is made with a full appreciation of irrational and inexplicable feelings as well as data.

Jewish Law and Transgenic Medicine

Jewish law — thousands of years old, constantly subject to debate and interpretation, but vested with the force of obligation in traditional Jewish communities — is recorded in a vast set of texts that begin with the Five Books of Moses and continue through three thousand years of commentary. I am not an expert on either the written nor the oral law. Only the equally ancient tradition of permitting disagreement for the sake of Heaven and the great good fortune of having found a small group of deep and pious teachers in the past few years has given me the courage to proceed with my own interpretations here.

The Jewish tradition has not yet reached a strong consensus on whether human germ-line gene enhancement — the technology of transgenic embryology that produced the green-eyed frog — has any possible use in our species. My own recent and brief immersion in the relevant Jewish texts has led me to some conclusions of my own, which are, of course, binding on no one but myself. DNA-based medicine is the paradigm of a wholly modern technology, one that the redactors of Jewish law could not have imagined a thousand years ago. Then as now, though, Jewish law welcomes new technologies, so long as they are not put to purposes that override current rabbinic interpretations of Torah.

So, for instance, using various tools for distinguishing the DNA of one person from another in order to identify a person is permitted, whether that person is dead or alive. A clear result from such analysis may even serve in the place of the traditionally required two witnesses in proceedings intended to fix responsibility for a death. But the purpose of the examination must not be to create an avoidable mess: DNA analysis may be used to identify the remains of a person who dies in an airplane crash, for instance, but it may not be used to determine whether or not a person is the biological child of both its legal parents, because there is a Jewish obligation to avoid seeking out illegitimacy.

In the Torah humans are defined as the progeny of Adam and Eve. In the Mishnah, the first books of the oral law, a human being is a child born of a human mother. Thus when the thought-experiment of a cow giving birth to a pig raises the question of whether that pig would be kosher, the consensus answer is yes, because its mother is kosher. Later commentators add human speech and human intelligence and, in particular, the capacity to distinguish good from evil, to the definition of a human being. These multiple definitions offer a range of possibility: a human may be narrowly defined as having all of these characteristics or more generously defined as having at least one.

The consensus seems to have fallen in the more generous direction, so for instance rabbinic commentators give a deaf-mute the same rights as anyone else, because he or she is born of a woman, and, in the famous Czech-Jewish thought-experiment, a golem cannot be put to death if it is intelligent, even though it was created without a woman's participation. Working from the generosity of the Jewish tradition in defining a person — either intelligent or born of a woman; one is sufficient — the technology of cloning a mammal from the nucleus of a tissue cell and the cytoplasm of an egg could in principle be extended to a human cell, a human egg, and, thereby, a human clone. From this perspective the germ-line modification of a human embryo would then seem to become a comparatively small perturbation of the ordinary course of events.

On the other hand, working from the primacy of free will in Judaism, the intentional modification of the genome of either sperm or egg — whether by cloning or by gene replacement, and whether intended to forestall a disease or for any other reason — would create a situation little different from slavery. If a child is in any sense the property of its parents, that is only because we presume them to be bound together by that most irrational of feelings, love. The person emerging from an experiment in which even one gene had been modified or replaced in all its cells would be the object of fascinated attention — if not the property — of the scientists and doctors who initiated his or her novel genome, and their funders. Their interest in that child would be in its experiences as an experiment, an interest hardly based on love; parental consent would merely legitimize a degree of disinterested ownership over another person from birth through death.

A cloned human would be a terrible experiment, performed on a person for his or her entire life, with no chance of that person withdrawing from the experiment if it does not go well. This is not the case for any medical treatment based on the modification of the genomes of tissue cells in a patient already among us. DNA-based gene-replacement therapies for such mutational diseases as cystic fibrosis and muscular dystrophy are not different from any other therapy designed to extend or improve a patient's life. Nor should rejection of a strategy to transgenically modify the human germ-line necessarily interfere with research aimed at producing novel drugs and hormones in the tissues of transgenetic animals or plants bearing human genes. Rabbi Gerard Wolpe, the director of the Louis Finklestein Institute for Religious and Social Science Studies at the Jewish Theological Seminary of America, makes these distinctions clear in a recent paper:

The fact that cloning offers the possibility of new clinical responses to disease makes it less of a moral threat. Cloning has been used as effective in treating such afflictions as hemophilia and cystic fibrosis. Once cloned, since all of the clones will have identical genetic characteristics, the cloned animal will be able to produce the same drug. Uses of cloning that could be life-enhancing persuaded authoritative Jewish bioethicists to give their approval at congressional hearings.

Choosing to Be Obligated

Religious obligations like these cannot be meaningful or compelling as such to someone who holds that it is not only irrational but trivial to argue a medical question from the point of view of an unknowable God's wish that we each have free will. My religious position that a human being should not be created for experimental purposes, even once — so far from any position one might take on data-based grounds, so deeply rooted as it is in the mysterious notion of the perceived wishes of a wholly unknowable God — turns out to have surprising popular support not only among religious Jews but also among many people of many religions and many of no acknowledged religion whatsoever.

Those who reject religious arguments such as mine often resist the experiment on the wholly secular ground that such a person, if born, would somehow reduce everyone else's humanity to that of a commodity, making everyone else somewhat less indispensable in their genetic uniqueness. Others are simply repelled by the notion of being in a genetic experiment that can only end with one's death.

These two "reasonable" reasons to resist the genetic modification of a human embryo — like other reasons that are not overtly religious — turn out to be eruptions of irrational feelings as well, feelings of unacknowledged discomfort with the facts of natural selection and evolution.

In terms of evolution's mutation-driven mechanisms, every one of us is precisely a genetic experiment that can end only in death. Our species' survival does not at all depend on any single person's life. As individuals, we are wholly expendable, as our ancestors were, human and before, since the beginning of life. Without the fact of mortality and the certainty that every individual of every species must die, there cannot have been the replacement of one set of forms and functions by another, the slow weeding and seeding of natural selection from which both we and frogs are current outcomes.

To resist a line of research because it reminds one of a fact of nature that makes one queasy is a good example of denial. Denial has its uses, but resistance to the data cannot be the source of a "rational" ethics. It is not consistent for someone who is irrationally trying to deny an aspect of nature, then, to argue that a religious source for a moral answer is irrational, as well, and therefore meaningless. There are indeed no data in science to justify the deep sense many of us share that making a flourescent green-eyed baby would be wrong. Without a religious context, only the wish to avoid thinking about the data stands between the new technology and the green-eyed baby. The religious, transcendent conferral of meaning and purpose on the uniqueness of any one person, and on that person's freedom to choose her fate, is of critical importance today precisely because its irrational freedom from all data of science liberates it from the burdens of denial and allows it to stand in the presence of all the data of evolution, painful as they may be to contemplate.

The Medical Uses of Religious Practice

Medicine and religion — whether two overlapping memes, or two responses to a call from beyond the knowable, or both — demonstrably overlap in their capacity to effect healing. In a recent series of research articles Harold Koenig of Duke Medical Center and his colleagues there and at Harvard Medical School have shown that persons given to regular religious observance have better outcomes in a variety of medical situations. These findings have met with a range of critical responses — mostly questioning the statistical validity of one or another study — but, at last accounting, the preponderance of the data support this surprising finding. As the *New Republic* put it a few months ago in an article about Dr. Koenig's work, with regard to any mainstream faith, "lack of religious involvement has an effect on mortality that is equivalent to 40 years of smoking one pack of cigarettes per day."

How can this be? Why should the antiselective memes of medicine and religion overlap, or, in my own particular religious view, why should the giving of meaning to life through medicine be in keeping with the revelations that tell Jews how our unknowable God wishes us to live? As we look at the overlap of medicine and my own religion it will be important to keep in mind that the insights articulated may be especially useful to Christians — and Muslims — who share the experience of different revelations of the same unknowable God. As Koenig points out, "Between Judaism and Christianity there is very little difference" in the effect of religious observance on delaying mortality.

Doctor's Orders Versus a Patient's Free Will

The shared Bible describes three different worlds, three successive experiments, of a sort, each of them acknowledging greater imperfection in human behavior. In Eden humans are immortal, vegetarian, and at peace. Outside Eden, where Cain kills Abel, they are mortal and violent. After the Flood they are no less violent, and, beginning with Noah himself, they are under the sway of uncontrollable sexual urges as well. Abraham emerges from this troubled world, and so do the three faiths that bear his name. The tribes who become the people Israel, living in this world, became the progenitors of a people for whom sickness and health were matters of God's dispensation. What is the place of medicine in such a world?

By way of an answer, here is a prayer I say every morning. The prayer was composed about sixteen hundred years ago by the Babylonian rabbi Abayei. For all that time its explicit purpose has been to make a blessing — that is, to give thanks to God — for one's physical integrity, immediately after having relieved oneself. It is even posted outside the bathrooms of many restaurants catering to observant Jews. I will use Dr. Ken Prager's translation of the earliest version we have in writing, from Talmud Brachot 6ob:

Abayei said when one comes out of the privy he should say "Blessed is He who had formed man in wisdom and created in him many orifices and many cavities. It is obvious and known before Your throne of glory that if one of them were to be ruptured or one of them were to be blocked, it would be impossible for a man to survive and stand before you. Blessed are You that heals all flesh and does wonders."

In the context of this prayer doctors and medicines are a redundancy at best and a blasphemy at worst. Why should anyone conversant with this universe of belief wish for another person to take the risk of arrogating God's power to heal flesh?

The first textual reference to what we would call a medical condition is after the Flood, in Genesis 8:21, where God acknowledges the inescapable presence of repressed memories that motivate self-destructive and aggressive behavior in later years: "Never again will I doom the Earth because of man, since the devisings of man's mind are evil from his youth." But when medicine itself first appears in Exodus 15:25-26 the context is unexpected and troubling to the modern eye: medicine should never be necessary except to compensate for moral failure, as disease is the result of a failure to follow God's commandments.

Right after the Jews escape from Egypt — at the cost of a God-sent plague that selectively kills all Egyptian firstborn, followed by a God-sent drowning of all their fathers — they find themselves in a waterless desert. After fixing the problem in a temporary fashion, God

put them to the test. He said "If you will heed the Lord your God diligently, doing what is upright in his sight, giving ear to His commandments and keeping all His laws, I will not bring upon you any of the diseases that I brought upon the Egyptians, for I the Lord am your healer."

Three thousand years later it is safe to say that most religious people, Jews and non-Jews alike, do not see medicine this way but rather as an aspect of our willingness and ability to assume a role our ancestors were commanded to leave to God. Just as we take on the capacity to change other aspects of the natural world through science — think of the seeding of clouds for rain — we assume permission to act in imitation of God by curing illness and enhancing fertility through medical interventions.

This is a new notion, not wholeheartedly endorsed by the judgments of traditional Judaism. The arrogation of an unknowable God's prerogatives of life and death brings with it the requirement that the patient surrender free will and the right to choose a future. To greatly oversimplify, Talmud counsels differently: the patient must be the one to choose to be treated and ought not choose to give away that free will choice, even to a doctor. The Talmud argues in many places that medicine, and by extension certainly the science that provides medicine with many tools, ought not to be given full authority over a sick person. Rather, the patient should be allowed to follow his or her intuition on the matter. A doctor's intervention is justified only insofar as a sick person seeks it out.

As Rabbi Yaakov Neuberger points out, this view is articulated most dramatically by the ruling in Talmud, Yoma 83a, that allows a patient to eat on the fast of Yom Kippur if he feels his life would be endangered by fasting, despite a doctor's certain assurances to the contrary. Doctors do receive the encouragement and approval of the Talmud, but this biblical dispensation is given precisely because it is understood that medicine will always be imperfect. The point is explicitly made that if medicine were certain of its cures, it would not need the dispensation in the first place.

This cautious, modest approach to the task of healing a sick person seems jarring and out of place in a scientific context. Seen from the world of basic research, where patients are submitted by lottery to double-blinded randomized studies — let alone from the cost-effective world of for-profit managed care — the notion that the patient must remain the arbiter of his or her fate seems quaint, even a bit perverse. The issue raised by the Jewish tradition here is not a matter of the data, that is, of medical science itself, but of its presentation to a patient, refracted through the lens of medicine. The modern physician's manner of presenting a diagnosis or taking a history often tends to carry with it a level of certainty and authority quite incompatible with the medical modesty recommended by Jewish tradition.

This is not solely a matter of overstepping by science, nor of bad manners among scientists, nor even of bad manners among doctors or their failure to think profoundly enough about the dignity, self-determination, and insight of a patient. There is as well a deep wish on the part on many people to surrender control of their situation to some authority, and illness can only add to any person's baseline of innate passivity.

When the wish to surrender authority over ones' self is no longer satisfied by prayer, science will do as well as any other authority. This may not be the normative Jewish way to see one's obligation to preserve freedom of action in the face of worldly authority, but it, nevertheless, has wide acceptance even in the Jewish world. As Rabbi Adin Steinsaltz has written, Jews are believers and the children of believers, who, when we cease to believe in God, continue at least to believe in the *New York Times*. So, for example, the theory of natural selection is taught in Israeli schools today as "Torat Darwin," a term that carries with it greater force than any scientific statement, of however great generality, deserves.

An Agenda for Medicine, from the Jewish Tradition

It remains difficult for me to overcome a lifetime of rational certainty and accept the possibility that my free will can be an expression of an unknowable God's hidden intentions, and that I must not surrender it to anyone, even my physician. Neither do I think it is possible any longer simply to paper over the issue and leave scientific medicine alone to fill the gap in moral authority. Without accepting its moral obligation to be concerned for the individual intentions of each patient, intentions that can be learned only by attention to each person's idiosyncratic past as well as their objective signs and subjective symptoms, a doctor will continue to be at risk of defaulting on her responsibility to a patient, regardless of treatment outcome.

The requirement that free will be preserved places a paradoxical limitation not just on the application of science to medicine but on all human interventions. In his essay "Catharsis" Rabbi Joseph Soloveitchik describes the paradox in a way familiar to secular scholars of the absurd: the essential element of heroism in Jewish terms is retreat. The paradigm of the hero who retreats is the patriarch Jacob, who wrestles an angel to the ground and then, instead of consummating his victory, lets him go. From such a withholding of final victory Jacob's descendants — the Jews of today — draw their continued existence:

The Torah wants man, who is bold and adventurous in his quest for opportunities, to act heroically, and at the final moment, when it appears to him that victory is within reach, to stop short, turn around, and retreat. At the most exalted moment of triumph and fulfillment man must forego the ecstasy of victory and take defeat at his own hands. . . . By freeing the defeated enemy Jacob defeated himself. He withdrew from a position he had won through courage and fortitude. He engaged in the movement of recoil.

To a person guided by Torah as the revelation of an unknowable but caring God, successful medical intervention need not confer any moral grandeur, nor need medical failure imply moral decay. From this Jewish tradition medicine can perhaps learn to recoil at the moment of victory, to pull back from the opportunity to take on the inappropriate role of judge of another person's fate.

It will be difficult to change the habits of the day, because for anyone — Jewish or not — who doubts there is an unknowable deity concerned for the moral content of their individual actions, medicine does have a way of filling the gap, sapping a patient's freedom to choose how to live his or her life. The enormous capacity of science to create tools for the manipulation of the natural world has helped confer moral authority on medicine by default. The lesson to be drawn from the Jewish tradition is that doctors and scientists have a moral obligation not to fill this gap with their own certainty, if for no other reason than to avoid losing their own God-given free will in turn.

How might this lesson from Judaism play out in more general operational terms in today's medicine? In the most general terms Judaism instructs us to redefine medical practice in the following ways. The profession — from basic researcher to primary-care physician — would accept the full autonomy of the patient at all times; the profession would see this autonomy of the patient as a critical aspect of the patient's identity as a unique and complete human being, regardless of physical or mental condition, and these obligations would not be less compelling in the last moments before a patient's inevitable death.

The mind/body dichotomy, and the current separation of mental and physical ailments, would have to be seen as senseless and damaging to the integrity of patient and doctor alike: the body is all. Indeed it is curious that any serious scientist still holds on to the notion of the mental as anything but physical; 150 years ago Darwin correctly called the mind an excretion of the brain, and today we can see it at work in the switching on and off of connections among the cells of the central nervous system.

Treatment of mental illness would be seen as medicine, pure and simple, and it would be possible to revisit aspects of medical care now suppressed because they fall at the boundary of mind and body, a boundary that we now know to be crossed at every instant, in both directions. In the same way, certain aspects of genetic medicine now emphasized as the wave of the future would be hedged about with precautions, as it would be understood that they risk the integrity of the patient, even as they reveal a great deal of prognostic information.

On the other hand, rogue aspects of medicine now suppressed because they are not easily reduced to testable models — psychosomatics and the placebo effect, for instance — might emerge as powerful tools for medicine. The notion of a psychosomatic effect should not be controversial, as we have all experienced one or more: they are the changes in the objective symptoms of an illness as a result of an induced change in self-image or mood. They can go in either direction; there are psychosomatic illnesses and there are psychosomatic cures. For myself, I know that calling my doctor with a set of symptoms makes those symptoms less troublesome, just as I know that imagining what these symptoms might imply instead of calling to find out makes them immediately worse.

In terms of the biology of the body, such psychosomatic effects are evidence that the mind and body are one, or, more precisely, that mental states are states of the body, occurring as the tissues of consciousness in the brain and the perceiving organs draw information from the body's physical condition and change that condition in turn. That much is safe to say on the basis of all we have learned about the brain in the past few decades.

Our medicine is nevertheless still embedded in a prior assumption, one that has remained unchanged since Aristotle. That is the notion that mental states are ineffable, existing somehow independent of the physical body, floating free in the matrix of the brain, unable to interact with any material part of the body. By holding on to that classic Greek presumption, modern medical practice falls unwittingly into a dogma wholly at odds with its own data.

The failure of modern medical practice to examine the consequences of holding on to this position continues to reduce all psychosomatic effects to self-deceptions, since they cannot be real: if the mind is immaterial, it cannot have material effects. This prior dismissal of psychosomatic events has cost the medical profession a good deal of opportunity to meet its Hippocratic Oath, as these effects do no harm, cost no money, and often bring about precisely the healing that medicine is sworn to seek.

Placebo Effects: The Utility of the Irrational

Of the many psychosomatic effects now in the shadows, I will highlight only one, the so-called placebo effect. *Placebo* means "I will please." A placebo is any objectively inert thing — a sugar pill, a word, a gesture — that has the effect of making someone feel better. A placebo's inertness makes it uninteresting to anyone for whom medicine is restricted to the intervention by active compounds and procedures that produce measurable and reproducible changes in the body. This same inertness makes a placebo interesting to anyone for whom the mind's workings are an example of the body at work, and important to anyone for whom the sole justifiable reason for any treatment is the improvement of the patient's condition.

A model of the mind as separate in substance from the body

is not so different from the data-free religious notion of an implanted, ineffable soul. The placebo effect is problematic in religious terms only for people who mistakenly conflate the ineffability and immateriality of an immortal soul with the material reality of mental states as expressions of the body's nervous tissues. Setting that confusion aside, certainly from a Jewish perspective and hopefully from a Christian one as well, clinical placebo effects should be welcomed, studied, and used.

One example of a placebo effect observed in a clinical setting, taken from an article in *Scientific American* by Walter Brown, will illustrate how powerful it can be and how important to rescue it from the limbo of our current ignorance of the chemical mechanisms by which it works. In a study of two hundred patients with physical complaints but no identifiable disease, doctors at the University of Southampton in England told some that no serious disease had been found and that they would soon be well; others heard that the cause of their ailment was unclear. Both statements were incomplete truths. Two weeks later 64 percent of the first group, but only 39 percent of the second group, had recovered. By extension, "bedside manner" — all the ways in which a physician presents herself to a patient — should make a difference in that patient's course, independent of the role of chemicals and agents that do make objective changes in the patient's tissues. Of all these bedside interventions, the one studied least and employed most is undoubtedly prayer.

Placebo Effects in Medical Research

Clinical placebo effects have always been known to occur in medical practice; they first got their current bad name not in the clinic but in the development of a technique called the double-blinded controlled experiment, which has in the past fifty years become the sole acceptable experimental setup for determining the effectiveness of any new drug or technique. In such a study neither the patient nor the physician knows whether the patient has received the drug or technique being tested or a sham compound or operation instead.

Fifty years ago, when the word *placebo* began to be used for the counterfeit treatment or pill that would allow a double-blinded study of a new drug, many doctors who had been using placebos in their practice objected. They knew that placebos in the first sense of the word — medicines given more to please than to benefit the patient — often had both effects, and they wished to keep the word for those cases. Confusion and ambiguity won out, though, and alternatives suggested, like *dummy,* did not prevail.

In analyzing the results of a double-blinded study, the effect of the dummy placebo is subtracted from the effect observed in the patients given the "real" treatment, and the leftover activity, if any, reveals whether the real treatment had any "real" effect at all. The problem with this way of analyzing the data is that it defines in advance all effects of the sham pill or operation as useless and meaningless. There is no a priori reason to do so except the presumption that the other kind of placebo effect, the clinical one that we all do experience, is also somehow not real at all.

The classic paper describing the ambivalent meaning of a strong placebo effect came out at the dawn of the age of scientific medicine, in 1945, in the *Journal of the American Medical Association*. The Harvard physician Henry Beecher reported that more than a third of patients suffering from pain, high blood pressure, asthma, or cough experienced some relief after taking a placebo.

Another study, by a contemporary of Beecher's, performed a sham operation — an incision in the chest wall — as a control for an operation thought then to assist in easing the pain of angina, the tying off of the internal mammary artery. Arterial ligation surgery is no longer done, but it was effective in 76 percent of the sixteen patients who received the operation. Remarkably enough, all — 100 percent — of the patients receiving the placebo sham operation also got better. As people with angina rarely get better if left alone, this result implies that the sham operation somehow helped resolve the problem — a true placebo effect — and that the placebo effect was sufficient to explain any positive result from the complete operation.

When double-blinded placebo-controlled studies were new, their creators saw such strong placebo effects as a real problem, because they raised the bar for efficacy of the compounds they wished to test. Their solution was to include in their studies a true control of doing nothing at all as well as a placebo control. Quite often there was a big positive difference between administering a sham drug or operation and doing nothing, in other words, a strong placebo effect. Remarkably, such effects were routinely but irrationally disregarded as getting in the way of "real" new treatments instead of being seen as the source of new treatments themselves. This may not have made sense, but it was accepted medical practice then, as it is today.

Current research on new drugs and procedures still sees the strong placebo as a problem to be overcome, not an aspect of treatment. As a result there remains a conflict between the goals of basic medical science — to suppress the efficacy of placebos in its randomized placebo-controlled testing of any "real" drug or procedure — and the moral obligation of medicine to give the patient the choice to ameliorate pain and suffering by any safe means at hand.

People vary in their response to a placebo, both one from another and over time for any one person. In a lovely paper in *Lancet* Henry McQuay reviews a number of double-blinded placebo-controlled studies on the effects of various compounds in relieving the pain of diabetic neuropathy. The placebo effect varied from paper to paper, from a low of 0 percent to a high of 80 percent. The effectiveness of the real compound also varied from study to study, from a low of 50 percent to a high of 90 percent. Remarkably, there was an excellent correlation in these studies between a high effectiveness of the compound tested and a high placebo effect as well. The reasons for such results are no more well understood than the reasons for the strong response when it does occur; our ignorance is the avoidable result of our reticence to study the strong placebo effect in its own right.

The scientific literature is thus rich in paradox. Just to pick one more example from the many published each year, adolescents suffering from "treatment-resistant" major depression were placed in a test where they might be given either what was then a new drug — the first-generation tricyclic antidepressant amitriptyline — or a placebo pill. Both groups showed approximately 70 percent to 80 percent improvement in clinical outcome measurements, and 65 percent to 70 percent in each group showed functional improvement as well.

Despite these remarkably good results, some brought on for an unknown reason simply by being enrolled in this study, the authors concluded that these youngsters were nonresponsive, as there was no difference between placebo and drug. Certainly under the Jewish assumption that a patient must be the arbiter of his or her treatment, this result would have led to attempts to capture the strong placebo effect for the benefit of future patients.

In an extreme attempt to avoid dealing with the placebo effect, some studies censor out those persons who show the strongest response to placebos. This is called a "placebo-washout," and the fact that it can be reported with a straight face tells us just how threatening a placebo is and how important it will be to have its effects studied and understood. This current practice may be extreme, but it does capture the irony of a medical profession that has so internalized a belief it cannot prove that it cannot make use of its own incompatible discoveries. Medical science must begin to deal with the gap the placebo effect creates between presumption and data, and religious figures must come to terms with the same gap from the other direction.

The Placebo Effect of Free Will: The Patient's Influence

The placebo effect is a natural subject for collaborative study among physicians and religious figures. When prayer works, it may well be a real placebo effect. Prayer as a placebo effect may not be a wholly welcome notion to someone who sees answered prayer as evidence for direct communication between an unknowable God and an ineffable soul. The Center for the Study of Science and Religion hopes to initiate a series of meetings among physicians and religious figures to engage the difficult task of building mutual understanding and cooperation between those who apprehend the existence of their immaterial soul, and those for whom the existence of the soul is not so important as the reality of the placebo effect.

Taking a careful look at the placebo effect will also mean taking a careful look at the most important aspect of medical care, the relationships between doctor and patient. The integration of psychosomatic phenomena like the placebo effect into medical practice could serve as a basis for overcoming the current uses of medical language to set doctor and patient apart. As Dr. Rita Charon has written, "Doctors differ from patients in the ways they use language and the purposes to which they put words. Doctors use words to contain, to control, to enclose. . . . Patients use language to express the sensations of things being amiss."

Acknowledging the reality of psychosomatic effects would help a doctor to hear a patients' use of language, obliging them to speak in that language in order to be effective. This could dramatically relieve the pain caused by missed communication, especially if the physician finds she is able to help the patient fix a part of what is amiss, when he is made to feel understood and appreciated as well as examined.

The current secular relationship, which encourages the doctor to allow the data of her tests to justify her becoming the sole arbiter of the patient's situation, is unsuitable for any serious study of the placebo effect. As the German historian of medical ethics Nikola Billar wrote,

The placebo effect is inherent in medical practice, but it is medicine's choice to neglect or employ its power. . . . If we do not reject the placebo effect as mockery, it can serve as a mirror, a chance for reflecting medicine's means as well as its goals. The ability to integrate the placebo effect in an ethically and medically adequate way could be a major achievement of modern medicine, which would benefit not only patients, but physicians as well, serving as a reminder of what lies at the heart of medicine: "one person treating another."

In this chapter I have tried to explain how, when an idea or a discovery of science is used by medicine to cure a disease, that idea acquires a new nonscientific meaning, the capacity to give back free will to a person who has lost it to illness. In this sense, good medicine is as much a religious as a scientific enterprise. DNA-based medical diagnosis is in a situation described by Pope John Paul II in the encyclical *Faith and Reason*:

All too soon . . . and often in an unforeseeable way this manifold activity of man . . . turns against man himself, at least in part, through the indirect consequences of its effects returning on himself. . . . Man therefore lives increasingly in fear. He is afraid [that] what he produces — not all of it, of course, or even most of it, but part and precisely that part that contains a special share of his genius and initiative — can radically turn against himself.

Everyone's DNA is different, and in those differences lie information about everyone's future health and life expectancy. As that information has winkled out of the genome it has presented medical science with another question that cannot be answered, except from the perspective of right and wrong: Is it right or wrong to tell someone what their fate will be — when they will die and what they will die of — if you cannot at the same time do anything to help them ward off that fate? That will be the question for the remaining chapter.

chapter three **Meaning Beyond Order:**
The Science of One Life at a Time

> Nothing that is worth doing is completed in our lifetime,
> therefore we must be saved by hope. Nothing true or beau-
> tiful or good makes complete sense in any immediate con-
> text of history; therefore we must be saved by faith. Noth-
> ing we do, however virtuous, can be accomplished alone;
> therefore we are saved by love.
>
> — Reinhold Niebuhr, *The Irony of American History*

I T IS HARD to live in two worlds. I would like to be able to say that every day I choose by my own free will to live my life according to the laws of my religion, but the truth is I do not. Instead I often find myself choosing reason over irrational obligation and cutting the corners of my religious obligations to myself and others. I would like to say, as well, that every day I find the strength to reopen my own examination of the natural world, through my eyes or the eyes of my scientific colleagues, ready to accept the implications of these discoveries no matter how hard they may be to fit into the rules by which I choose to live. But here too I often fall short, walling myself off from these implications in order to focus on the beautiful, elegant details of nature.

There are days when both traditions cooperate, and the workings of nature fit well enough with what for the moment I sense are an unknowable God's intentions, that I can both understand an aspect of the world's design and feel its purpose, and the meaning it gives to my place in the world, at the same time. On the other days — most days — I feel I must pick between data and feeling. That is just the choice genetic medicine seems to face today. In both cases this choice is no choice at all: for my state of mind and for genetic medicine's current practices, the denial of feelings for the sake of the data is not a moral option. The real choice made available by free will is always to work with both the feeling and the data, or with neither. On my bad days, as in many of genetic medicine's current practices, the choice is not to use one or the other, but to use neither.

The better choice for medicine, as for me, is always to use both. In medicine the most important field for this choice to be made more widely available, right away, is the field of medical diagnosis informed by human genetics, where the tools of science offer the opportunity to obtain data that a proper regard for feelings would oblige us to leave in darkness. Brought to light in the wrong context, these data must cause a vast amount of bad feeling and unnecessary emotional pain. There are many examples to choose from in describing this situation. I will discuss in detail one I am most intimately familiar with: the genetic markers — the specific DNA sequences found in the genomes — of some but not all Jews.

Medical Prognosis and the Genetic Bottleneck: The Dangers of a Little Knowledge

Genetic medicine is the branch of science that depends on a knowledge of many lifetimes, much history, and vast collaboration. The trick will be to see that it is informed, as Niebuhr would have it, by hope, faith, and love. Certain current practices in genetic medicine do not promise much of any of these three, all of them being the irrational properties of a religion.

In my first two chapters I discussed memes — ideas with a

life of their own — and my hope that our free will would allow us to choose such anti-Darwinian religious memes as hope, faith, and love. In this last portion of the book I will consider ways that hope, faith, and love might be returned to genetic medicine, that part of medical practice invested in a future of ever expanding genetic knowledge. Without our choice to act now to return to them, every person will one day be obliged to pay attention to genomic news in a context possibly not devoid of faith or love but certainly devoid of hope.

There already are a few people who have had to deal with genetic news in this gloomy context. These are the descendants of genetic bottlenecks, members of groups of people — apparently unrelated — who share a small number of common ancestors. The Jews of Eastern Europe — the Ashkenazim — are one of these groups. How Jews and others respond to this challenge should be of interest to everyone whose recent family history includes the inheritance of unusual versions of one or more genes, that is, everyone.

Human Diversity with Genetic Certainty?

The importance of each person's individual choices is central to the Jewish tradition, but it is nevertheless easy enough to lose sight of free will in the details of obligatory observance. The worst mistake one can make in this regard is to think one has understood the meaning of a ritual when one has, in fact, missed the point completely. A little more than a year ago, for example, I received a notice by e-mail telling me of a departmental faculty meeting that would fall on the first day of the Jewish new year, Rosh Hashanah. I sent back an e-mail saying, No, I'd be celebrating the 5,758th anniversary of the creation of the universe that day, and could not be at the meeting.

My absence annoyed many members of my department; some who were Jewish thought I was kidding, others who were not thought I was just nuts. I thought I had acted honestly, freeing myself for this celebration of creation. I had forgotten that the day has other names: Yom HaZikharon, or Memorial Day, and Yom HaDin, the Day of Judgment.

Afterward, when I told Rabbi Steinsaltz of my colleagues' response, he asked, could my department perhaps have been trying to teach me? Were my colleagues perhaps annoyed at me not for my piety but for my ignorance? Rosh Hashanah, the first day of the Jewish year, is not the anniversary of the creation of the world. It is the anniversary of creation's sixth day, the hours of the appearance of our common human ancestors, their short stay in Eden, and their exile into a world where their further actions — freely chosen — would have consequences.

The ancient Jewish recognition of a shared ancestry of all people produces a second, equally unquestioning presumption, one that emerges from the idea of a Day of Judgment: that the unmeasurable, infinite value of each human life derives not from any aspiration to perfection, but precisely from the inherited differences that allow each of us to look different, and to choose differently, from all others.

The earliest part of the oral law, or Talmud — the Mishnah — is a record of expectations and laws binding on Jews, codified almost two millennia ago. When I first began to speak on the utility of my own tradition to my science some years ago, Professor David Weiss-Halivny gave me a reference in Mishnah Sanhedrin, which has a commentary on the book of Genesis that makes this point with special elegance. The Mishnah is giving the reasons why witnesses to capital crimes must be taught that a person's life is at stake in their testimony, and that any person's life is a more serious matter than most anything else. The Mishnah then comments:

For this reason one individual Human Being was created . . . to proclaim the greatness of the Holy One the Blessed: for a man strikes many coins from one mold and they all resemble one another, but the supreme King of Kings, the Holy One the Blessed, stamped every human in the stamp of the first human being and yet not one of them is like the other. Therefore every person is obliged to say: the world was created for my sake.

What remains today of the certainty that "yet not one of them is like the other?" At the deepest level of the letters in our DNA genomes, it is indeed the case that "not one is like the other." Except for twins and other children who emerge from the same single fertilization of an egg by a sperm, any two people reading this book will have genomes three billion letters long that differ by about one letter in every few hundred.

Because the genome is so wonderfully long, even siblings have genomes that differ in millions of places. Genetic variation among parents and the iron rule of sexual reproduction — that in the production of sperm or egg a choice will be made for every gene, with one version discarded and one version passed on to the next generation — guarantee that while children of the same parents may resemble each other they will not be identical unless they come from the same fertilized egg.

This raises an interesting question: if we are all so different from one another, why do siblings resemble each other more than any two people chosen at random? Brothers and sisters — and even cousins, who share grandparents rather than parents — look similar, even though each is genetically distinct from the others, because their only genetic differences are taken from a very small number of choices, the particular versions of any given gene carried in their parents' genomes.

Even a three-generation family is genetically restricted by the versions of genes available from common grandparents, although the restriction is moderated for cousins by the genetic choices provided to each through a different second set of grandparents. That is why the resemblance among cousins is usually less striking than among siblings but far greater than among two randomly chosen people. The broad generalization that people look more different from one another the less they are related by recent common ancestry also tells us that the human species has been, over most of its history, quite happy to make babies with strangers.

If our species' history was instead one of widespread inbreeding or endogamy — the selection of a mate from one's own extended family rather than from a strange one — then we would expect that, over time, different extended families — and therefore sets of parents within those extended families — would each carry the badge of their history as a set of shared versions of the human genome. The data we have tell us instead that, over the past few dozens of centuries since the last ice age, human genomes have been constantly sifted and resifted in the making of babies, with very little in the way of ancestral fastidiousness beyond the grandparental level to reduce the genetic variation from one extended family to another.

Each of us is a member of a very recent family that shares only a tiny fraction of the total human genetic diversity available. We are nevertheless all members of one family in the deep historic sense that our species has interbred widely for most of its history; that is why there are no versions of a gene that are present only in one place on the planet, and none that are wholly absent from any reasonably large population, no matter how isolated.

In between the nuclear family of the greeting card and the

global family of the Mishnah, there are a host of other, less recognized subpopulations, or "families," of various sizes, each made up of people sharing the genetic choices of their distant common ancestors. Genetic diseases are sometimes said to run in families. Which families do we mean? In fact, inheritance of the propensity to develop a disease can occur in all three sorts of families.

Families and Diseases

Versions of genes associated with a disease may have survived in a family or a subpopulation for many reasons. Some may be here as the result of natural selection, good examples of the cold difference between its power to select for a difference that aids the survival of a species and its complete disinterest in any one individual's fate. One could imagine the utility to a species that is always hungry of any version of a gene that left an individual alive and well long enough to make many babies and that then — by speeding mortality — removed him or her from the pool of superfluous persons still needing to be fed.

Other nonfunctional versions of a gene might be present in a population because natural selection has exerted no grip on them, or because the variant that causes a disease in modern times might have been one that conferred a survival advantage in earlier, less hygienic times. For instance, well-fed overweight people descended from the survivors of many generations of near-starvation often show a "thrifty" metabolism, a survival response to caloric restriction that makes weight loss difficult. Natural selection is wonderfully obtuse, and what might have been necessary for survival under one set of circumstances can easily become a burden under another.

The life expectancy of the human species in the absence of

sufficient food, clean water, and separate sewage disposal is about half of what it is in the presence of such nineteenth-century medical advances. As life expectancy increases for more of the human population, we are bound to discover more such versions of genes in our midst. Any versions of a gene that might have helped to survive starvation, parasites, or fecal contamination of food and water would have been strongly selective until only a short time ago. If they also led to death in the fifth decade or later, no one would have even known about that until very recently.

Although the evidence for past utility must remain circumstantial, we have some good examples of this. Mutations in red-blood cell hemoglobin that may cause sickle-cell anemia or Thalassemia are with us today because they offer strong survival value in the presence of malaria, and mutations in a gene that can cause cystic fibrosis at one extreme may well be with us today because they mitigated the lethal consequences of cholera in the past.

Founder-Effect Families

Two historical explanations of a late-onset inherited disease — the random noise of neutral mutation or the palimpsest of earlier advantage — fit well into the agenda that gives medical meaning to more immediate genetic conditions: isolate the gene, understand how it works in the normal case, and provide treatments to ameliorate its mutational disability. But there is a third historical explanation for the persistence of a genetic malady in a subpopulation, one producing circumstances for doctor and patient alike that open the risk of a breach in the capacity of medicine to confer a moral meaning on DNA-based data.

That breach occurs when the reason for inheritance of the condition is not drawn from the deep past, nor from recent family history, but from the mid-range. In those cases, having inherited a gene associated with a disease may be the consequence of the history of nations, and DNA-based diagnosis of it will not easily be extricated from its political and religious implications.

Consider the cheetah: as a species in the wild it has been threatened for centuries, its coat much sought after, its habitats encroached upon, its behavior poorly compatible with human society. Almost obliterated more than once in the past, all cheetahs today are the descendants of a very few ancient cheetah families. As a result, today's cheetahs — apparently unrelated, certainly not able to mate with each other when they are raised in separate zoos — are nevertheless all far more alike in appearance than any two randomly picked lions or tigers. Their DNA tells why.

For any given cheetah gene there is a very good chance that any two cheetahs share the same version, because all but a very few versions of any gene were lost when all but a very few ancestors were killed in past encounters with our species. Cheetahs are not quite clones of one another, but they are as alike — or more so — than a set of children in one large family. Survivors of a genetic bottleneck, cheetahs that appear to be unrelated have DNA that tells us they are survivors of a past disaster, the genetic constriction called a founder effect.

A high probability of inheriting precisely the sort of version of a gene that no one wishes can be the result of having an unexpected founder effect in one's past family history. The founding members of such a family would have had to survive a great cataclysm, and after the fall there would have been no alternative but to marry one's relations.

Descendants of these founding families would have no choice but to bear the consequences of a purely accidental selection of the founders' versions of genes. A fateful version, one that gets trapped in a founder population simply because the founding family happened to have carried it, will not be shed from that family so long as strangers cannot become spouses.

By itself, a disaster that leaves only a small subpopulation alive usually does not assure a large family of people sharing the risky gene, because trapped genes will be dispersed into a larger population each time a family member chooses a mate who is not descended from the founding families. But when the surviving subpopulation has previously defined itself by rules that forbid such marriages to outsiders, then the problem of founder-effect inheritance may become severe. When all three requirements are met — a small founder population trapping one or more deleterious versions of genes, a strict adherence to endogamy for many generations, and great fertility during that long period — there will then be a significant population of people carrying the same deleterious mutation for this historical reason.

While the human species as a whole is not the product of any detectable founder effect more recent than its emergence as a species in Africa some millions of years ago — that is the meaning of the typically enormous variability of human genomes from one person to the next — there are many populations across the planet whose history makes them unexpectedly, invisibly, more like cheetahs than they know. For a person born into a founder-effect population the data of DNA-based medicine may have specific but medically irrelevant meanings: DNA-analysis may lead to the recovery of family relationships that transcend the borders of language and appearance and that mark one out in helpful or dangerous ways.

Membership in a group that defines itself by its behavior may thus become presumptive membership in a genetically marked population. For people in such a situation, any medical meaning of DNA-based diagnosis is shadowed by this fact. If the group shares versions of genes that are deleterious, and if the group has habits that maintain genetic isolation from the general population, then membership in the group may be seen by outsiders as tantamount to a disease in and of itself.

At such times it is the duty of medicine to reassert its initial meaning over the science that reveals the situation and to protect such populations from serious potential nonmedical consequences of a visit to the doctor.

Even though a founder-effect descendant's DNA may contain useful scientific data — and the DNA of such people is much sought after for research purposes — it should not be too easily given away. DNA-based information about a member of such a subpopulation can be medically meaningful only if the person at risk can be safe in the choice to find it out. If the information is obtained in confidence, and is explained solely and completely in the context of the person's future risk, with comprehensive and sensitive counseling on the possibilities of future problems, then it may be able to revert to medically useful information.

Even then it is doubtful that there is a clear medical use for information about one's DNA-driven fate when there is nothing to do to deflect that fate. At such times DNA-based diagnostic information may have a greater meaning in political than medical terms. DNA differences associated with disease set their carriers aside as a separate population for such political purposes as insurance rating, government or private employment risk or military service, or worthiness of receiving support for an expensive education.

Finally, such DNA-based information may have meanings within the endogamous founder population itself, revealing just how well or poorly the rules of endogamy have in fact been followed. All such purposes for reading a person's DNA cannot possibly make the acquisition of DNA-based information about the members of a group medically — and thereby emotionally and religiously — meaningful. To confuse or elide the difference between a medical meaning and any of these others is to betray one of the moral obligations of medicine, at great potential cost to oneself, one's friends, and one's family.

Perhaps the best example of how easy it was in the last decade of the twentieth century — as easy as it was in the fourth — to use genetic "medicine" as a cover for punishment is in the People's Republic of China. As recently as 1994 the provinces of Liaoning and Gansu were still enforcing laws mandating forced sterilization and abortion in families with a history of retinal degeneration and "autosomal disease on one side of a couple." Decoding this law, it means sterilization of someone who has inherited a damaged version of a gene from one parent and a functional version from the other parent. Such people are not sick in any way, nor will they be; every one of us is the bearer of just such mixtures of one good and one bad version, for hundreds of our genes. As *Nature Genetics* put it in a 1997 editorial: "The observations of Chen Minzhang, the Minister of Public Health, are not reassuring: apparently births of 'inferior quality' are serious among 'the old revolutionary base, ethnic minorities, the frontier and economically poor areas.'"

Are Jews a Family?

It would be a mistake to think that misuses or abuses of diagnostic techniques for distinguishing one person's DNA from another's are a problem only for members of a founder population such as my own. Once again, in this chapter I ask the indulgence of you all, as I use my own ancestors to illustrate a broad issue — in this case the risk that genetic information may have inadvertent, punitive uses. Thanks to the very facts of our species-wide shared genetic heritage, so clearly revealed at the DNA level, this issue is of deep relevance to everyone who thinks seriously about their future and their family's future, regardless of their religious preference, their degree of piety, or their scientific expertise.

We Jews call ourselves a family, and in many ways — for better and worse — act as if we were. We continue to preserve common laws, habits, languages, texts, and historical memories as well as the belief that all these are the gifts of an unknowable Deity who began our place in history by exchanging covenental promises with three successive generations of our ancestors, Abraham's, Isaac's, and Jacob's.

We have preserved these shared habits and beliefs for millennia, over a large fraction of the populated world. They are on the one hand the very model of strong memes in action, sometimes symbiotic, sometimes not; on the other, they exemplify the durability and reality of religious belief in the unknowable and the survival of belief in the face of millennia of strong negative selective pressure. Do these facts mean that the Jews of today are a biological family as well, linked by descent from shared ancestors? Yes and no.

The Jews of centuries ago who codified prayers understood that while being born a Jew was precious and important, it was not necessary and it certainly was not sufficient. The central ideas and actions of a Jew have always had to be taught and learned; they have never been inherited. Nevertheless, until about a decade ago many reasonable people could still make the argument — in the absence of evidence to the contrary — that since Jews accept the covenants made with Abraham, Isaac, and Jacob, the genomes of Jews must somehow be different from the genomes of all other people, containing unique versions of many genes; that is, that Jews are a biological family.

The difference between "Are Jews a family?" and "Do Jews all share the same versions of one or more genes?" is that the second form of the question has a testable, precise answer. As no two people have exactly the same version of the human genomic text, this claim could be confirmed or rejected by a search for versions of the human genome shared by all Jews and no other people.

Unfortunately, the first group of scientists and doctors to pose the question in this way did so in a scientific context that reduced people to the bearers of their genomes, and in an inhumane manner that wound up being so painful, so cruel, and so lethal that it is difficult to ask it again, even two generations after they were finally put out of business.

This context was the Nazi notion that, despite all appearances to the contrary, every potential Jewish parent was inevitably the bearer of an undesirable alien inheritance that would crush the true inheritance of Germany; in other words, the same idea that we have been asking about, the notion that Jews are a biological family. However inarticulately stated by Hitler's propagandists, this was the scientifically certified argument for the destruction by bullet, gas, and fire of German and then European Jewry, of Germans and others who had one Jewish grandparent, and of about a million Jewish children, some of them born when I was born, in 1940.

The Demographic Consequences of a Founder Effect: the Ashkenazic Jewish Example

In this historical context it is all the more remarkable that Jews all over the world flock to the new technology of DNA-based diagnosis, eager to lend their individual genomes — each a surviving data point from the terrible experiment in negative selection — to a revisiting of this issue of biological Judaism. Fortunately, this self-absorbed curiosity has provided sufficient genetic material to give a perfectly clear negative answer: there is no support in the genomes of today's Jews for the calumnious and calamitous model of biological Judaism. There are no DNA sequences common to all Jews and absent from all non-Jews; there is nothing in the human genome that makes or diagnoses a person as a Jew. But, as often happens when the tools of science are used in a medical context without a medical purpose, these same studies have raised unexpected difficulties in both medical and religious contexts.

Everyone will have to face similar difficulties sooner or later. To see what is coming in everyone's future, and to begin to understand how the Jewish experience may be helpful in heading off the worst possible outcomes for everyone else, it is necessary to know the demographics of the Jewish past. Numbers of Jews at various times in the past are not easy to get, and the ones we have are not precise. The best estimates I have been able to find show that, in the long view, Jewish populations have enjoyed only a few periods of smooth uninterrupted growth in any one part of the world.

From the earliest records to the admittedly partial documentation we have of numbers of Jews from different countries in the past five hundred years, a repeated pattern emerges of relatively brief periods of rapid population growth followed by instances of severe, almost complete population collapse, with long intervening periods of low but stable numbers. Each time a group of Jews survived one of these boom-and-bust cycles, it was as the descendants of a very small number: symbolically, as with the Jewish Patriarchs and

Matriarchs, from just one family. Today there are about 6 billion people on Earth, and about 13 million of them are Jews. This means that about one person in five hundred, worldwide, is Jewish.

Three thousand years ago, upon David's establishment of the first Jewish nation-state, the Jewish population worldwide rose from about five hundred thousand to about 2 million, and for a while almost one person in fifty, worldwide, was Jewish. But once Jerusalem fell to Babylon twenty-five hundred years ago, Jewish numbers declined to fewer than three hundred thousand, and afterward, under Greek rule, the Jewish fraction of the population was once again reduced to about one in five hundred.

Numbers rose to about 4 or 5 million about two thousand years ago. They remained high, and for the first two Roman centuries Jews were as many as one person in sixty. Then, as the world's population grew, the Jewish fraction fell once again, to about one person in two hundred by the year 600. For the next millennium, until about 1600, the number of Jews remained about 1-1.5 million while the world's population more than doubled. As a result, the fraction of the world that was Jewish kept shrinking, until by 1600 it was back to the one in five hundred that it was during the Babylonian exile.

Then an unexpected thing happened: a boom occurred in a fertile part of Europe, and it did not go bust for almost four hundred years. The Pale — a part of eastern central Europe now partially contained within the borders of Russia, Belarus, Ukraine, and Poland — let Jews live. The medieval Hebrew name for these European lands was Ashkenaz, and so Jews from that region still refer to themselves by the Hebrew plural Ashkenazim. Ironically, the place name derives from the name of one of the grandsons of Noah by his son Japheth, which suggests that, from the beginning, local inhabitants of this region were understood by their Jewish neighbors to be very distant relations, indeed, not even the descendants of Noah's son Shem, the ancestor of Abraham.

For four hundred years the Jews of Ashkenaz stayed put and grew in numbers. In that period — 1500 to 1900 — the total number of Jews worldwide went from about 1 million to about 11 million, and almost all of that increase took place among the Ashkenazim. By 1900 1 person in 150 worldwide was Jewish, and more than 90 percent of this greatest number ever were descended from the Jews who had been living in central Europe since the 1500s. Nineteen thirty-nine was the peak year for Jews in this world, who numbered between 16–17 million and were about 1 person in 120 worldwide. All but about a million of them were descended from the original settlers of Ashkenaz, although the total number of Jews living there actually decreased in the early twentieth century because of the emigration that brought my grandparents, among many others, to these shores.

The demographic losses of 1939–1945 may not be recoverable. In the last thirty years Jews worldwide have numbered about 13 million, neither growing nor shrinking by much. As the world's population booms, the Jewish slice of it shrinks. That is why today the fraction of the world's people who are Jewish — one in five hundred or less — is no higher than it was twenty-five hundred years ago, after the first Babylonian exile.

My point in reviewing this set of population figures is not to raise the question of why, since David's kingdom fell, the world has been unable to bear more than one Jew in every five hundred people for longer than a century or so. Just the opposite: three times — after the fall of David and Solomon's kingdom twenty-five hundred years ago, after the fall of the Hasmonean kingdom two thousand years ago, and in Ashkenaz after the pogroms and the crusades and the Black Death of the middle ages five hundred years ago — the total Jewish population actually grew, and in each case it grew more rapidly than the general population.

Of these three instances the startling growth of the Jewish population in Ashkenaz is also the source of one of the world's largest founder effects. In 1500 only a few percent — some tens of thousands — of the world's million or so Jews lived in the Ashkenazic Pale. By 1939 about 95 percent of the world's 17 million Jews either lived in Ashkenaz or were the descendants of people who had lived there until no more than fifty years earlier.

The Medical Consequences of a Founder Effect

Combining the history of Ashkenaz with data from the genomes of their descendants alive today, we can get a good estimate of how few families founded today's Ashkenazic Jewish population. When people who carry an inherited condition are also the descendants of a single ancestor, their versions of the affected gene will be identical. If in addition they are the descendants of a population that practiced endogamy, then they will share more of their genome than that one gene with others suffering the condition.

Given the great number of versions of each gene available in the human species at large, long runs of identical versions of genes in two unrelated people will never occur by coincidence. But because the surviving population in Ashkenaz was so terribly small in the mid-1600s, and because it grew in an uninterrupted way from such small numbers, a large fraction of Jews today share such long stretches of genes with each other.

This was known in principle, but nevertheless the discovery a few years ago of identical stretches of DNA hundreds of genes long, in hundreds of apparently unrelated people from all corners of the world, was a surprise. The people who offered their genomes for this landmark study shared only two things: an inherited tendency to have one's muscles twisting one about — called Idiopathic Torsion Dystonia (ITD) — and an ancestor who came from Ashkenaz.

Most people in this study, but not all, called themselves Jews. Sometimes, though, members of an affected family would be shocked to discover that the inherited condition that had brought them into the study very likely meant an unexpected Jewish ancestry. With surprising regularity, when they understood the meaning of the tests done on themselves and their children, they would remember, admit, but not always accept, having Jewish ancestors.

The data from this study argued very strongly that the oddities of fate and the murderous intentions of strangers had fixed a history of near extinction four hundred years ago in the DNAs of the majority of Jews alive today. According to the scientists who carried out this study, the utter sameness of DNA in persons inheriting ITD worldwide means that every Jew whose ancestors come from Ashkenaz — about nine of every ten Jews alive today — is the descendant of one of no more than about three thousand families who survived the pogroms of the mid-1600s.

Diseases of the Askenazic Bottleneck Are Not Jewish Diseases

It is terribly sad that Jews allow these marks of history to be called "Jewish diseases." As the Chief Rabbi of London once famously said in response to an article in the *London Times* about the early-onset, lethal, incurable neurological condition called Tay-Sachs Disease, "There are no Jewish diseases," only the past consequences of violent anti-Semitism. Clearly, the shared genes of the Ashkenazim do not define any aspect of their Jewishness. Those descended from Ashkenazic ancestors share a higher-than-average frequency of versions of various genes only because they are descended from the same survivors of Jewish Ashkenaz. The genomes of other Jews reflect their different histories. Descent from an Ashkenazic family, with or without its attendant inherited conditions, cannot make a person Jewish.

Those who see any aspect of Judaism as inherited must be ignoring the demonstrated fact that Ashkenazim are a founder-effect family genetically quite different from the non-Ashkenazic families who make up most of the Jews of Israel. These Israelis would certainly fail any biological criterion set by Ashkenazic history, and vice versa. Equally clearly, shared genes bring a shared fate: those Jews who do share a common Ashkenazic ancestry may not have inherited their Jewishness that way, but many have inherited a shared fate in the form of a genetic problem. Setting the particularities of Jewish history aside, then, let us look at a particular gene associated with an all-too-common late-onset disease, and at some of the medical and religious implications of what has been recently learned about both.

"Genetic disease" has as many meanings as "family." The diversity of our species tells us that many different versions of each gene can be compatible with a healthy life. An unknown number of other variant genes are wholly incompatible with embryonic development; inheritance of any of them leads to the loss of an embryo before birth.

In between are the variants of a gene that are compatible

with birth but not with the birthright we have come to expect, a life expectancy of around eighty years or more. Some of these variants are active in early life, causing infantile or childhood inherited diseases like Tay-Sachs. Other variant genes are not called upon by the body for much of a person's lifetime.

Still other variants may lack functionality but have no immediate consequence, because a second copy of the gene is able to carry the work for both. When that sole copy of a functional gene is lost — and the older we get the more chance there is of a copy of a gene getting lost in one of our cells — the absence of a second functioning gene may show itself as a late-onset inherited disease, like inherited breast cancer.

DNA analysis can be used to find affected individuals in any of these different sorts of families. There are some good clinical reasons for seeking out these genes and the people who carry them. Once the affected gene is recovered, it can be used to find the functional version from another person, and with that in hand the search for understanding how the normal gene works becomes straightforward science. Also, a working version may be used to repair the tissue damage caused by a nonfunctional gene. For instance, victims of cystic fibrosis who lack a fully functional version of the gene called CFTR have had their symptoms alleviated, at least for a time, by the administration of large doses of DNA encoding a functional human CFTR gene. On the other hand, members of families with a history of a late-onset inherited condition may find themselves obliged by the same technology to learn about their fate from their genomes at a time when they have no symptoms nor any expectation of treatment once the symptoms appear.

Cancer of the breast will afflict about one person in nine in this country. Each tumor begins as a mutation in a breast tissue cell. When an error in copying hits a critical gene, a cell is freed from the restrictions of differentiation and it begins to grow, forming a clone of mutated cells. Free growth is often accompanied by a loosening of a cell's editorial proofreading capacity, so that one mutation may beget another. In time a clone becomes a bump, the bump becomes a lump, and the lump becomes a spreading, dangerous tumor. About 75 percent of the cases of breast cancer are sporadic. At this moment these three-quarters of cases begin with mutations that have no known specific cause, whether genetic, hormonal, or environmental. A third of the remaining quarter of cases — about one in seven cases overall — occur in families with three or more generations of victims. Such families have inherited a clear susceptibility to the disease, and their tumors may well be the result of having inherited one or more nonfunctional variant genes. The causes of the remaining 15 percent of cases are ambiguous, because neither family histories nor environmental histories are sufficiently clear to determine the likelihood of a preexisting genetic risk or shared exposure to a known carcinogen.

BRCA1 and Breast Cancer: An Askenazic Case Study

BRCA1 (BReast CAncer 1) is the first gene whose variants were found to be associated with familial breast cancer. Mutations in this gene were found by University of Utah scientists in breast cancer victims from a number of unrelated Mormon families whose members suffered from a high incidence of breast cancer for three or more generations. BRCA1-associated cancer was reproducibly different from the more common sporadic type: in addition to occurring in families, it appeared at a very young age, in both breasts, and in association with other cancers, especially of the ovaries. In such families the link between cancer and the mutation is very high indeed: victims had an 86 percent chance of carrying a mutation in BRCA1. BRCA1 proved to be a very long gene, with lots of room for mistakes. More than two hundred different mutations of BRCA1 have been recovered from different high-incidence families in the decade since the gene was first isolated and sequenced.

Ashkenazic Jewish families with a history of breast and ovarian cancer may also inherit a mutation in BRCA1. The founder effect of Ashkenazic history predicts that Ashkenazic families should have inherited only a few of the many known mutations in that gene; so far all Ashkenazic families with a history of breast cancer who do have a familial mutation in BRCA1 have been found to carry either one or another of only two of the two hundred known mutations in BRCA1.

This simplicity makes Ashkenazic families in general a population of great interest both to scientists working on the details of how BRCA1 works in its normal and mutant states and to genetic epidemiologists interested in setting up large-scale scans for single mutations. There has been a remarkable willingness on the part of Ashkenazic families — those who do not suffer from generations of breast cancer as well as those who do — to assist both these branches of science.

Not all the work these scientists wish to do, though, has any medical content. For instance, a 1997 paper in the *New England Journal of Medicine* reports a study in which scientists obtained the cooperation of hundreds of members of Ashkenazic families who did not have a family history of breast cancer and checked their genomes for mutations in BRCA1. The results were a disturbing, unexpected, and unwanted prophetic revelation through science: 2 to 3 percent were carrying one of the two mutations.

This meant that even in the absence of any members with symptoms among relatives of two or three generations, and certainly in the absence of any symptoms in oneself, everyone in one of these families had a much higher risk of a bad fate than other people — even other Jews — who were not living out a founder effect laid down by the violence of their ancestors' enemies. Statistically, each woman in a breast cancer-free Ashkenazic family who is found to carry one of the two mutations in BRCA1 has a greater than 50 percent chance of developing a breast tumor and a 20 percent chance of developing an ovarian tumor.

What is to be done with these prophecies? They do not come to families prepared by a prior history of disease for news of an inherited condition. Rather, they come to healthy people — from unaffected families — on the wings of ancient and recent history, reminders that we all are not only the descendants of our grandparents but also of their ancestors, people with whom we may think we have nothing in common.

We can be sure that prophetic news of this type — an unclear but high risk of a dread disease at a time when there are no family or personal symptoms — will not be reserved solely for Ashkenazic Jewish families in the future. The difficulties lie not only in the discovery of a problem when you had no idea you were at risk; they lie in the wish to do the right thing, when there is no clear idea what that would be.

In the case of a BRCA1 gene there are only three options once it is discovered that one carries a mutation, and in the absence of a family history of the disease none alone justify submitting one's DNA to find out. Ovarian surgery means early menopause as well as sterility; prophylactic breast removal is a major operation and while it does lower risk, it does not remove it completely; and surveillance for the appearance of a breast tumor is something every woman should be doing anyway.

Making Sense of a BRCA₁ DNA Diagnosis

When a physician takes the family history of a patient, a great deal of information may emerge that is not of immediate relevance to the condition that brings the patient in but that will help provide the context for understanding the patient's needs and for helping the patient decide what course of action to take. Studies of the actual utility of DNA testing in susceptible subpopulations show that the information obtained from a DNA sample is all but useless in the absence of a knowledge of family history. The problem is that DNA-based data is much simpler to obtain than to understand. This is a problem for both patient and doctor. For the patient, the issue is simple: proper counseling. As a recent paper in the *New England Journal of Medicine* put it, in reporting on DNA testing of a few hundred people of various ages, all of whom had good reason from family history to suspect a mutation accompanied by colon polyps and cancer,

Only 18.6 percent received genetic counseling before the test, and only 16.9 percent provided written informed consent. In 31.6 percent of the cases physicians misinterpreted the test results. . . . Physicians should be prepared to offer genetic counseling if they order genetic tests.

For the physician, the problem is simple as well: doctors in primary care and psychiatrists who provide counseling to people at genetic risk need to know more about genetics and genetic tests. In a 1993 paper, a group led by Neil Holtzman surveyed thirty-two hundred physicians with a questionnaire designed to assay the demographics of their practice, their knowledge of genetics, and their awareness of the availability of genetic tests.

More than half responded. Their average grade was 74 percent of the knowledge items, compared with 95 percent for genetics professionals given the same questionnaire. This is not a very good average — certainly it is far beneath the grades these doctors got as premedical students — but remediation is possible. Along with obvious predictors of new knowledge like recency of graduation from medical school, high scores correlated with "practicing . . . [where] exposure to genetics problems is likely," and "not using pharmaceutical companies as a source of information about new medical practices." For a physician, remediation requires, in other words, knowing your patient and knowing your stuff.

Criteria to screen for BRCA1 mutations are in flux, but already we can see the outlines of a religiously informed sensitive policy, one drawn from the experiences of the Askenazic community but not limited to that community. Here, for instance, are the guidelines used by Dr. Freya Schnabel, a Columbia colleague whose practice includes many Ashkenazic families. First, the screen is not to be made available to everyone: in the general population the grounds for a BRCA1 test are either three cases of breast cancer in the family, two in women under sixty. In Ashkenazic families the criteria are slightly less restrictive but still stringent: at least two cases of breast cancer, at least one in a woman younger than sixty.

Second — not a criterion but a boundary — prenatal screening for this adult-onset disease is not to be performed even in affected families. As there is no way to predict the age of onset or whether the child will indeed develop a tumor, there are no grounds to put forward the choice of abortion as there might be with inherited diseases that occur with complete certainty at birth or at a very early age.

The third, fourth, and fifth criteria speak to the essential inseparability of mind and body and the central importance of psychosomatic and emotional events. Third, anyone entering the process of DNA diagnosis for BRCA1 status must first be counseled and a judgment made of their ability to understand and accept either a positive or negative result; fourth, only people who are able to accept the lack of clarity of either result, and are willing to make decisions for themselves about the consequences, nevertheless, should be assayed; fifth, counseling, both psychological and genetic, should continue for at least as long after the result is in as before the decision to be tested.

We may not all be members of high-risk families, but we are all at risk. It remains uncertain whether guidelines like these — which go far toward helping people to choose how to approach the future, while not unnecessarily shadowing anyone's free will with useless genomic determinism — will be applied widely or whether genomic data stripped of any medical meaning will continue to be imposed on us all.

Biological Judaism: Bad Science, Bad Religion

I have argued that meaning and purpose are necessary, and that medicine as well as religion can be a source of meaning for the data of science. But even in religions with a long history of endogamy, the use of DNA data to make claims of inherited religious sensibility is inherently wrong. When those claims overlap medical issues, they allow for an extremely dangerous confusion to reemerge from the ashes of history.

When medicine confuses religious faith with biological ancestry and science links biological ancestry to genetic difficulty, then it is but a small slippery step downward for medical practice to mark out member of a religious group as genetically defective per se. The logic will be familiar, as will the threat of it, but this time the tools are available to uncover evidence of common ancestry, and of common genetic difficulties, in any population, worldwide.

Two immediate issues arise from the power of DNA analysis to uncover ancient common ancestries. One pertains to Israelis and their neighbors, and one to everyone. I might as well get to the first through a passage from Torah, because it is a really tough issue. From Genesis 25, lines 7–9: "This was the total span of Abraham's life: one hundred and seventy-five years. And Abraham breathed his last, dying at a good ripe age, old and contented; and he was gathered to his kin. His sons Isaac and Ishmael buried him in the cave of Machpelah."

It should not come as a surprise to learn — recall — that exiled Ishmael, circumcised patriarch of the twelve Arab tribes, rejoined his half-brother, the Jewish patriarch Isaac, to give their father a proper burial alongside Isaac's mother. If the tradition of descent from Isaac links Jews together despite the absence of biological confirmation for that ancestry, it must also link Jews forever with their Arab cousins.

One day the Jews of Israel will have to ask whether there can be an Israeli Law of Return that makes sense while excluding the children of Ishmael. The only two countries that have a "law of return" are Israel and Germany. In both countries, and in no others, a person born outside the country may receive citizenship on request by virtue of religion, or "blood," while other persons born inside the borders may not receive citizenship, for similar reasons. Israelis and Germans cannot tell each other how they should see their past, but Israelis certainly ought to think about why they are still in this tiny club.

Sacred Diversity: The Religious
Meaning of Common Ancestry

The second implication of DNA-based diagnosis is for everyone to ponder: people — our species — are one family in precisely the same way that Jews are not. The story of Ashkenazic inherited diseases should make us all sensitive to the larger issues of inherited disease, and of genetic difference. Beyond the obligation this story urges us all to undertake — to accept the evidence and give up vain hopes of any religious birthright in their genes — is an even larger moral duty.

The moral context that gives meaning to science through medicine requires the attention of both science and medicine to a person in all his or her complexity and variability. The linkage of scientific medicine to religious history rather than to religious values may be more interesting in scientific terms, but it is fatally dangerous in medical terms.

Perhaps the best way to see the difference is to understand that though in social terms people tend to aggregate into groups of majority and minority populations — often separated by religion — by the data of our genomes we are all members of genetic minorities that range in size from the millions of a founder population, to the dozens of an immediate family, to the irreducible minority of one that is at the heart and soul of medicine. We would do well to acknowledge that nothing in the legacy of human DNA blocks the choice to value the differences among us above the resemblance any of us might have to our idea of an ideal person.

With all of us exerting our free will to decide whether it is wise to know more or to know less about our own DNA at any given moment, the scientific data of DNA-based medicine may be returned to a proper medical context. In light of the DNA evidence we already have, this means stretching the definition of normal variation to include the greatest possible diversity of inherited appearances and behaviors. Our obligation here is as clear, in its own way, as is the countervailing trend in current medical science.

The straightforward agenda of scientists and the short-term acquiescence of physicians fifty years ago led to the creation of the National Institutes of Health, each institute named for a disease of the middle-aged white men in Congress who gave out federal money in those days and still do today. These institutes have provided the country and the world with much knowledge of great value, both medical and monetary. But with the creation of cheap, easy scans for mutations in genes like BRCA1, knowledge contributed by NIH-supported science has begun to change medicine in ways that deny the meaning medicine provides to science.

In *The Missing Moment* I drew the following quote from my mentor and teacher, James D. Watson, discoverer of the structure of DNA and founding director of the Human Genome Project. Writing in the annual report of his laboratory, he had said:

If we could use genetic analysis to help work out the biochemical pathways underlying memory and clear thinking, for example, we might be able to find pharmaceutical compounds to improve these most needed human attributes. Thus, those who want to protect the mentally ill or the slow learner may not get what they strive for if they portray them exclusively as victims of their environment. We might like to think otherwise, but only by reducing the differences in human beings will we ever have a society in which we can effectively view all individuals as truly equal.

I admire Jim Watson for his unmatched taste in picking the right question to ask of nature as much today as I did when I first met him in the late 1960s, but I know that here he was deeply wrong. We know from a century and a half of research in ecology and evolution that as a species our future lies not in minimizing our differences but in cherishing them. We know as well from millennia of religious insight that there is no possible way to justify any ranking of one person over another on grounds of any aspect of their physical being. From those two insights we have the chance of working toward a properly informed medicine, capable of using any and all insights from science in a context derived from the insights of many religions and thereby capable of reducing all data to one purpose: to help people in need, one person at a time.

The social and political structures of medical care today encourage neither physicians nor patients to seek these meanings from medicine. Rather, for reasons that are difficult to fathom from a religious perspective but easy enough to understand in fiscal terms, both will have to wrench themselves out of current practices if they are to help each other re-create a meaningful version of medicine.

Neither physicians nor patients can be free to make their choices, nor true to themselves, so long as they are both distracted by the constraints of politics and money. This is easy to say, but hard to fix. Hard, but not impossible; it is never impossible to choose to act well.

For example, Rita Charon, a professor of medicine at Columbia Presbyterian Hospital, has a practice of medicine relatively free of these constraints. Her patients — triply marginalized old, African American women — see her in a public clinic, a place so far beneath the fiscal radar as to leave them both free to discover themselves.

Dr. Charon has written of her changing relationship with each of her patients as she has come to know them and be known by them over decades of repeated visits to the clinic. As she has undoubtedly extended their lives by her practice of clear, rational scientific medicine, they have often changed her by sharing aspects of their lives that — on the face of it — have no clinical significance, and would be lost in any scientifically restricted, wholly rational form of medical practice.

Often the epiphanous moment that links science to religion is at the core of this reciprocal growth and flowering, as in this passage from one of her research papers:

An 82-year-old obese, diabetic, hypertensive woman with osteoarthritis has been in my practice for around fifteen years. Our early years together were marked by disagreement over silly things: she insisted on name brand medicines, even when generics were just as good, and I bristled at the extra work and cost. She, too, never took seriously the need to address her obesity. Consequently, her diabetes was ill-controlled and her degenerative knee disease disabling. One morning, as she sat on the examining table waiting for me to take her blood pressure (which was invariably alarmingly high and triggered in me anxiety, fear of reprisal, great impatience, and the felt duty to scold), she mentioned that she sang in the church choir. I don't know why, but I asked her to sing me a hymn. This woman, whose body habitus I routinely described as "morbidly obese" was transfigured into a form of stateliness and dignity as she raised her heavy head, clasped her hands, and sang in a deep dark alto about the Lord, on the banks of the river, bringing her home. From then on, I would do anything for her and she for me.

That is what medicine looks like when the doctor keeps all tools of science at her fingertips, when the meaning of those tools is given by the mysterious capacity for free will, and the choice to use them to preserve another person's life and health, and when the person who uses these tools for that purpose knows herself to be no different in any important way from her patient.

At such times, and in such places, medicine may give both doctor and patient the chance to share an experience of the Unknowable. So long as such times and places are forbidden by the rules of managed care, or by the habits of a science that uses the needs of medicine to pay for its research, medicine will suffer an unnecessary loss of meaning. So long as we passively acquiesce to these constraints, we will all continue to suffer from an avoidable loss of meaning in our own lives.

Postscript

I BEGAN MY PREFACE with a short remark by Auden and wish to end the book with a longer quote from his work, and a single lapidary sentence by Freud. Both remind us of the work we each have to do, to be prepared for either scientific insight or religious revelation: no one is less interesting than a lazy person imitating wisdom.

For the poem I am grateful to Professor Ed Mendelson. It is from W. H. Auden's Phi Beta Kappa poem for Harvard College, "Under Which Lyre," in which the rational forces of Apollo seem to be winning their war with the ecstatic devotees of Hermes. Auden, always a Hermetic, spoke these words in 1946, at the closing of World War II and the opening of the golden age of science:

> In our morale must lie our strength:
> So, that we may behold at length
> Routed Apollo's

Battalions melt away like fog,
Keep well the Hermetic Decalogue,
     Which runs as follows:
Thou shalt not do as the dean pleases,
Thou shalt not write thy doctor's thesis
     On education,
Thou shalt not worship projects nor
Shalt thou or thine bow down before
     Administration.
Thou shalt not answer questionnaires
Or quizzes upon World-Affairs,
     Nor with compliance
Take any test. Thou shalt not sit
With statisticians nor commit
     A social science.
Thou shalt not be on friendly terms
With guys in advertising firms,
     Nor speak with such
As read the Bible for its prose,
Nor, above all, make love to those
     Who wash too much.
Thou shalt not live within thy means
Nor on plain water and raw greens.
     If thou must choose
Between the chances, choose the odd:
Read "The New Yorker," trust in God;
     And take short views.

And, finally, the sentence from Freud. I have tried to say that the work of change begins with a willingness to change one's own presumptions. Freud put it this way: "An unacknowledged wish cannot be influenced." If this book has been as useful to read as it was to write, I think it can only be because in some way I have helped you to acknowledge a wish of your own.

Bibliography

Books of General Interest

Abram, David. 1997. *The Spell of the Sensuous*. New York: Vintage.

Barnavi, Eli, ed. 1992. *A Historical Atlas of the Jewish People*, pp. 1–2. New York: Schocken.

Bonne-Tamir, Batsheva, and Avinoam Adam, 1992. *Genetic Diversity Among Jews: Diseases and Markers at the DNA Level*. New York: Oxford.

Bynum, Caroline Walker. 1995. *The Resurrection of the Body*. New York: Columbia University Press.

Cahill, Thomas. 1998. *The Gifts of the Jews*. New York: Talese.

Cavalli-Sforza, Luigi Luca, and Francesco Cavalli-Sforza. 1995. *The Great Human Diasporas*. New York: Addison-Wesley.

Chomsky, Noam. 1993. *Language and Thought*. Wakefield, R.I.: Moyer Bell.

Coles, Robert. 1999. *The Secular Mind*. Princeton: Princeton University Press.

Cohen, I. Bernard, and Richard Westfall, 1995. *Newton: A Norton Critical Edition*. New York: Norton.

Damasio, Antonio. 1994. *Descartes' Error: Emotion, Reason, and the Human Brain*. New York: Putnam.

——. 1999. *The Feeling of What Happens: Body and Emotion in the Making of Consciousness*. New York: Harcourt Brace.

Darwin, Charles. 1985 [1859]. *The Origin of Species*. New York: Penguin.

Dawkins, Richard. 1976. *The Selfish Gene*. New York: Oxford.

Dorff, Elliot. 1998. *Matters of Life and Death*. Philadelphia: Jewish Publication Society.

Dworkin, Ronald. 1993. *Life's Dominion*. New York: Knopf.

Eliott, George. 1984 [1876]. *Daniel Deronda*. Ed. Graham Handley. Clarendon. Oxford: Oxford University Press.

Fortey, Richard. 1998. *Life: A Natural History of the First Four Billion Years of Life on Earth*. New York: Knopf.

Freedman, Harry. 1983. *Midrash Rabbah*. Trans. M. Simon. 3d ed. London: Soncino.

Gillman, Neil. 1997. *The Death of Death: Resurrection and Immortality in Jewish Thought*. Woodstock: Jewish Lights.

Gilman, Sander. 1988. *Disease and Representation: Images of Illness from Madness to AIDS*. Ithaca: Cornell University Press.

Himelstein, Shmuel, trans. 1994. *Mishnah Seder Neẓikin* Vol 2. Jerusalem: Maor Wallach.

Horgan, John. 1996. *The End of Science*. New York: Broadway.

James, William. 1985 [1902]. *The Varieties of Religious Experience: A Study in Human Nature*. New York: Penguin Classics.

Jamison, Kay Redfield. 1996. *An Unquiet Mind*. New York: Vintage.

Jewish Publication Society. 1999. *JPS Hebrew-English Tanakh: The Traditional Hebrew Text and New JPS Translation*. Jewish Publication Society: Philadelphia.

Jung, Leo, trans. 1974. *Hebrew-English Edition of the Talmud: Yoma*. London: Soncino.

Kitcher, Philip. 1996. *The Lives to Come*. New York: Simon and Schuster.

——. 1996. *Abusing Science: The Case Against Creationism*. Cambridge: MIT Press.

LeDoux, Joseph. 1996. *The Emotional Brain: The Mysterious Underpinnings of Emotional Life*. New York: Simon and Schuster.

Lown, Bernard. 1996. *The Lost Art of Healing*. Boston: Houghton Mifflin.

Lowontin, Richard. 1982. *Human Diversity*. New York: Scientific American.

Margulis, Lynn. 1982. *Early Life*. Boston: Science Books International.

Pollack, Robert. 1975. *Readings in Mammalian Cell Culture*. Cold Spring Harbor, N.Y.: Cold Spring Harbor Laboratory Press.

——. 1994. *Signs of Life: The Language and Meanings of DNA*. Boston: Houghton Mifflin.

——. 1999. *The Missing Moment: How the Unconscious Shapes Modern Science*. Boston: Houghton Mifflin.

Pope John Paul II. 1998. *Encyclical Letter Fides et Ratio (on the Relationship of Faith and Reason)*. http://www.cin.org/jp2/fides.html

Preuss, Jacob. 1993. *Biblical and Talmudic Medicine*. Trans. Fred Rosner. Northvale, N.J.: Aronson.

Proctor, Robert. 1988. *Racial Hygiene: Medicine Under the Nazis*. Cambridge: Harvard University Press.

Rosner, Fred. 1993. *Medicine and Jewish Law*. Vols. 1–2. Northvale, N.J.: Aronson.

Sarewitz, Daniel. *Frontiers of Illusion: Science, Technology, and the Politics of Progress*. Philadelphia: Temple University Press.

Scholem, Gershom. 1971. *The Messianic Idea in Judaism*. New York: Schocken.

Simon, Marcel, trans. 1960. *Hebrew-English Edition of the Talmud: Berakoth*. London: Soncino.

Sinclair, Daniel. 1989. *Tradition and the Biological Revolution: The Application of Jewish Law to the Critically Ill*. Edinburgh: Edinburgh University Press.

Soloveitchik, Joseph B. 1998 [1962]. *The Lonely Man of Faith*. Northvale, N.J.: Aronson.

Steinsaltz, Adin. 1988. *The Strife of the Spirit*. Northvale, N.J.: Aronson.

———. 1995. *On Being Free*. Northvale, N.J.: Aronson.

———. 1999. *Simple Words*. New York: Simon and Schuster.

Tipler, Frank. 1994. *The Physics of Immortality*. New York: Simon and Schuster.

Weiner, Jonathan. 1995. *The Beak of the Finch*. New York: Vintage.

Wolpert, Lewis. 1992. *The Unnatural Nature of Science*. London: Faber and Faber.

Yerushalmi, Yosef Hayim. 1982. *Zakhor*. Seattle: University of Washington Press.

Research Articles

Amaya, E., M. F. Offield, and R. M. Grainger. 1998. Frog genetics: Xenopus tropicalis jumps into the future. *Trends in Genetics* 14: 253–255.

Benedetti, F., M. Amanzio, S. Baldi, C. Casadio, and G. Maggi. 1999.

Inducing placebo respiratory depressant responses in humans via opioid receptors. *European Journal of Neuroscience* 11:625–631.

Benson, H., and R. Friedman. 1996. Harnessing the power of the placebo effect and renaming it "Remembered Wellness." *Annual Review of Medicine.* 47:193–199.

Biller, N. 1999. The placebo effect: mocking or mirroring medicine? *Perspectives in Biology and Medicine* 42:398–401.

Birmaher, B. et al. 1998. Randomized, controlled trial of Amitriptyline versus placebo for adolescents with "treatment resistant major depression." *Journal of the American Academy of Child Adolescent Psychiatry* 37:527–535.

Brody, H. 1997. Placebo response, sustained partnership, and emotional resilience in practice. *JABFP* 10:72–74.

Brown, W. 1998. The placebo effect. *Scientific American*, January, pp. 90–95.

Bynum, C. W. 1997. Wonder. *American Historical Review* 102:1.

Cassel, E. 1982. The nature of suffering and the goals of medicine. *New England Journal of Medicine.* 306:639–645.

Charon, R. 1992. To build a case: medical histories as traditions in conflict. *Literature and Medicine* 11:115–132.

Dandhayuthapani, S. et al. 1995. Green fluorescent protein as a marker for gene expression and cell biology of mycobacterial interactions with macrophages. *Molecular Microbiology* 17:901–912.

Dawkins, R. 1995. God's utility function. *Scientific American*, November, pp. 80–85.

DeMarco, C. W. 1998. On the impossibility of placebo effects in psychotherapy. *Philosophical Psychology* 11:207–227.

Dennett, D. C. 1995. Darwin's dangerous idea. *Sciences* May-June, p. 34.

Easterbrook, G. 1999. Is religion good for your health? Faith healers. *New Republic*, July 19/July 26, pp. 20–23.

Elander, G., and Hermeren, G. 1995. Placebo effect and randomized clinical trials. *Theoretical Medicine* 16:171–182.

Ernst, E., and K. L. Resch. 1995. Concept of true and perceived placebo effects. *British Medical Journal* 311:551–553.

Fan, S., J.-a. Wang, and R. Yuan et al. 1999. BRCA1 Inhibition of estrogen receptor signaling in transfected cells. *Science* 284:1354–1356.

Feit, C. 1990. Darwin and drash: the interplay of Torah and biology. *Torah U-Madda Journal* 2:25–36.

Flaten, M. A., T. Simonsen, and H. Olsen. 1999. Drug-related information generates placebo and nocebo responses that modify the drug response. *Psychosomatic Medicine* 61:250–255.

Foster, M. et al. 1997. Communal discourse as a supplement to informed consent for genetic research. *Nature Genetics* 17:277.

Freedman, B., K. C. Glass, and Weijer, C. 1996. Placebo orthodoxy in clinical research II: ethical, legal, and regulatory myths. *Journal of Law, Medicine, and Ethics* 24:252–259.

Garcia-Alonso, F., E. Guallar, O. M. Bakke, and X. Carne. 1998. Use and abuse of placebo in phase III trials. *European Journal of Clinical Pharmacology* 54:101–105.

Gaylin, W. 1994. Knowing good and doing good. *Hastings Center Report*, May-June, pp. 36–41.

Goldgar, D., and P. Reilly, 1995. A common BRCA1 mutation in the Ashkenazim. *Nature Genetics* 11:113.

Goodenough, O., and R. Dawkins, 1994. The "St. Jude" mind virus. *Nature* 371:23.

Goodwin, J. 1995. Culture and medicine: the influence of puritanism on American medical practice. *Perspectives in biology and medicine* 38:567.

Gould, S. J. 1999. Darwin's more stately mission. *Science* 284:2087.

Grazi, R. V., and J. B. Wolowelsky. 1992. Review: Preimplantation sex selection and genetic screening in contemporary Jewish law and ethics. *Journal of Assisted Reproduction and Genetics* 9:318–322.

—— 1993. Homologous artificial insemination (AHI) and gamete intrafallopian transfer (GIFT) in Roman Catholicism and halakhic Judaism. *International Journal of Fertility* 38:75–78

Greenberg, R. P., S. Fisher, and J. A. Ritter. 1995. Placebo washout is not a meaningful part of antidepressant drug trials. *Perceptual and Motor Skills* 81:688–690.

Greenwood, J. D. 1996. Freud's "Tally" argument, placebo control treatments, and the evaluation of psychotherapy. *Philosophy of Science* 63:605–621.

Hacia, J. et al. 1996. Detections of heterozygous mutations in BRCA1 using high density oligonucleotide arrays and two-color fluorescence analysis. *Nature Genetics* 14:441.

Halperin, Mordechai, 1993. Modern perspectives on halakha and medicine. In F. Rosner, *Medicine and Jewish Law*, p. 169. Northvale, N.J.: Aronson.

Harris, W. et al. 1999. A randomized, controlled trial on the effects of remote, intercessory prayer on outcomes in patients admitted to the coronary care unit. *Archives of Internal Medicine* 159:2273.

Hemila, H. 1996. Vitamin C, the placebo effect, and the common cold: a case study of how preconceptions influence the analysis of results. *Journal of Clinical Epidemiology* 49:1079–1084.

Hoffman, K. et al. 1993. Physicians' knowledge of genetics and genetic tests. *Academic Medicine* 68:625.

Hummer, R. 1999. Religious involvement and U.S. adult mortality. *Demography* 36:273–285.

Huxley, T. 1896. *Science and the Hebrew Tradition*. New York: Appleton.

Kaptchuk, T. J. 1998. Powerful placebo: the dark side of the randomised controlled trial. *Lancet* 351:1722–1725.

Kass, L. R. 1988. Evolution and the Bible: Genesis 1 revisited. *Commentary*, November, pp. 29–39.

—— 1999. The moral meaning of genetic technology. *Commentary*, September, pp. 32–38.

Kendler, K. et al. 1997. Religion, psychopathology, and substance use and abuse: a multimeasure, genetic-epidemiological study. *American Journal of Psychiatry* 154:322.

Koenig, H. et al. 1998. The relationship between religious activities and blood pressure in older adults. *International Journal of Psychiatry in Medicine* 28:189–213.

—— 1999a. Editorial: religion, spirituality, and medicine — a rebuttal to skeptics. *International Journal of Psychiatry in Medicine* 29:123–131.

—— 1999b. Does religious attendance prolong survival? A six-year follow-up study of 3,968 older adults. *Journal of Gerontology* 54A:M370.

Kronn, D., V. Jansen, and H. Ostrer, 1998. Carrier screening for Cystic Fibrosis, Gaucher disease, and Tay-Sachs disease in the Ashkenazi Jewish population. *Archives of Internal Medicine* 158: 777.

Li Wan Po, A., J. Pegg, and S. Bridges. 1995. Editorial: The placebo response — friend or foe? *Journal of Clinical Pharmacy and Therapuetics* 20:47–48.

Ludmerer, K. 1999. Instilling professionalism in medical education. *Journal of the American Medical Association* 282:881.

Martinson, J. J., N. H. Chapman, D. C. Rees, Y.-T. Liu, and J. B. Clegg. 1997. Global distribution of the CCR5 gene 32-basepair deletion. *Nature Genetics* 16:100–103.

Mintz, A. 1994. Words, meaning, and spirit: the Talmud in translation. *Torah U-Madda Journal* 5:115–155.

Motulsky, A. 1995. Jewish diseases and origins. *Nature Genetics* 9:99.

Nesse, R. M. and K. C. Berridge. 1997. Psychoactive drug use in evolutionary perspective. *Science* 278:63–65.

Neuberger, Y. 1991. Halakha and scientific method. *Torah U-Madda Journal* 3:82.

Oldham, J. A. 1996. Should we be clinically exploiting the power of the placebo effect? *Reviews in Clinical Gerontology* 6:213–214.

Ploghaus, A., I. Tracey, J. S. Gati, S. Clare, R. S. Menon, P. M. Matthews, J. N. P. Rawlins. 1999. Dissociating pain from its anticipation in the human brain. *Science* 284:1979–1981.

Posner, M. I., and M. K. Rothbart. 1998. Attention, self-regulation, and consciousness. *Philosophical Transactions of the Royal Society, London* 353:1915–1927.

Prager, K., 1997. For everything a blessing. *Journal of the American Medical Association* 277:1589.

Prasher, D. et al. 1992. Primary structure of the Aequorea victoria green-fluorescent protein. *Gene* 111:229–233.

Priebe, S., and M. Broker. 1997. Initial response to active drug and placebo predicts outcome of antidepressant treatment. *European Psychiatry* 12:28–33.

Rapp, R. 1988. Chromosomes and communication: the discourse of genetic counseling. *Medical Anthropology Quarterly* 2:143.

Reilly, P. et al. 1997. Ethical issues in genetic research: disclosure and informed consent. *Nature Genetics* 15:16.

Risch, N. et al. 1995. Genetic analysis of idiopathic torsion dystonia in Askenazi Jews and their recent descent from a small founder population. *Nature Genetics* 9:152; correspondence in *Nature Genetics* 11:15.

Rosner, F. 1998. Judaism, genetic screening, and genetic therapy. *Mount Sinai Journal of Medicine* 65:406–413.

Rubinsztein, D. et al. 1995. Mutational bias provides a model for the evolution of Huntington's Disease and predicts a general increase in disease prevalence. *Nature Genetics* 7:525.

Sato, A., C. O'Huigin, F. Figueroa, P. R. Grant, B. R. Grant, H. Tichy, and J. Klein. 1999. Phylogeny of Darwin's finches as revealed by mtDNA sequences. *Proceedings of the National Academy of Sciences* 96:5101–5106.

Sech, S. M., J. D. Montoya, P. A. Bernier, E. Barnboym, S. Brown, A. Gregory, and C. Roehrborn. 1998. The so-called placebo effect in benign prostatic hyperplasia treatment trials represents partially a conditional regression to the mean induced by censoring. *Urology* 51:242–250.

Selig, S. et al. 1997. Y chromosomes of Jewish priests. *Nature* 385:32.

Shattuck-Eidens, D. et al. 1997. BRCA1 sequence analysis in women at high risk for susceptibility mutations. *Journal of the American Medical Association* 278:1242.

Sher, L. 1997. The placebo effect on moood and behavior: the role of the endogenous opioid system. *Medical Hypotheses* 48:347–349.

Slack, J. M. W. 1996. High hops of transgenic frogs. *Nature* 383: 765–766

Sloan, R. et al. 1999. Religion, spirituality, and medicine. *Lancet* 353:664–667.

Soloveitchik, J. B. 1978. Majesty and humility. *Tradition: A Journal of Orthodox Thought* 17:25–37.

—— 1978. Catharsis. *Tradition: A Journal of Orthodox Thought* 17:39–54.

Spiro, H. M. 1998. The optimist. *Science and Medicine*, September/October, pp. 2–3.

Steinsaltz, A. 1995. Where do Torah and science clash? *Torah U-Madda Journal* 5:156–167.

Straus J. L., and Von Ammon Cavanaugh, S. 1996. Placebo Effects: issues for clinical practice in psychiatry and medicine. *Psychosomatics* 37:315–326.

Struewing, J. et al. 1995. The carrier frequency of the BRCA1 185delAG mutation is approximately 1 percent in Ashkenazi Jewish individuals. *Nature Genetics* 11:198.

Vogel, G. 1999. Frog is a prince of a new model organism. *Science* 285:25.

Volvaka, J., T. B. Cooper, E. M. Laska, and M. Meisner. 1996. Placebo washout in trials of antipsychotic drugs. *Schizophrenia Bulletin* 22:567–576.

Waldfogel, S., and S. Meadows, 1996. Religious issues in capacity evaluation. *General Hospital Psychiatry* 18:173.

Watson, J. D., and F. H. C. Crick. 1953. A molecular structure of nucleic acids: A structure for Deoxyribose Nucleic Acid. *Nature* 171:737–738.

Wilmut, I., A. E. Schnieke, J. McWhir, A. J. Kind, and K. H. S. Cambell. 1997. Viable offspring derived from fetal and adult mammalian cells. *Nature* 385:810–813.

Wolpe, G. I., The genome and the Jew. *Jewish Theological Seminary Magazine* 8:1999.

Zajicek, G. 1997. Editorial: A scientific framework for the placebo effect. *Cancer Journal* 10:236–237.

Zernicka-Goetz, M., J. Pines, K. Ryan, K. R. Siemering, J. Haseloff, M. J. Evans, and J. B. Gurdon. 1996. An indelible lineage marker for *Xenopus* using a mutated green fluorescent protein. *Development* 122:3719–3724.

Zygar, C. A., T. L. Cook Jr., and R. M. Grainger. 1998. Gene activation during early stages of lens induction in *Xenopus*.

Index

Gene(s): associated with disease, 77–78, 79, 80, 81, 82, 90–91; second copy of, 91; variants, 90–91

Gene replacement, 51–52

Gensis, 6, 56, 74, 98

Genetic bottlenecks, 79; Ashkenazic, 89–92; medical prognosis and, 72–73

Genetic counseling, 95, 96, 97

Genetic disease, 90–91

Genetic experiment, modification of human embryo, 48, 51, 53

Genetic information: punitive uses of, 83–84; when nothing can be done about it, 69, 81, 91, 94

Genetic markers, in Jews, 72, 73

Genetic medicine, 8–9, 61, 72–73; punitive uses of, 82

Genetic uniqueness, 53

Genetic variation, 75, 76–77, 99

Genetics, physicians' knowledge of, 96

Genome(s), 23, 29, 30, 41–42, 69, 75, 76, 99; of frog, 45; intentional modifications of, 51; of Jews, 83–84, 85, 88–89, 90; manipulation of, 47–48; variability of, 80

German physicians, 47–48

Germany, 98–99

Germ-line gene enhancement, 49, 51

Gifts of the Jews, The (Cahill), 21

God, 3, 5; in Jewish tradition, 34, 35, 40; power to heal, 56–57; putting science in place of, 4; scientific discourse and names of, 6–7

Goodenough, Oliver, 31

Gould, Steven J., 26–28

Green-eyed frog, 42–45, 49

Green Fluorescent Protein (GFP), 45

Group membership, 80–81

Harvard Medical School, 54

Hasmonean kingdom, 88

Hippocratic Oath, 62

Hitler, Adolf, 84

Holtzman, Neil, 95

Hope, 72, 73

Human behavior, imperfection in, 55

Human beings: definitions of, 50–51; differences among, 100, 101

Human diversity, with genetic certainty, 73–77

Human life, value of, 74–75

Human species, 46–47, 80; interbreeding, 76; life expectancy, 77–78

Huntington's chorea, 48

Huxley, T. 28

Hypotheses, 13, 14, 32

Ideas: competing, 29; natural selection of, 7; as parasitical memes, 30, 32; revealed by insight, 13–14, 15–16; revela-

McQuay, Henry, 66

Meaning, 2, 7–8, 37; choosing irrational path to, 32–34; in facts, 5; free will and, 8, 35, 36; ideas of, as memes, 30, 31; from insight, 16; lacking, in origin of species, 24, 25; loss of, 103; and medicine and transgenes, 42–45; of natural world, 28; necessity of, 97; beyond order, 71–103; in order, 6, 39–69; order versus, 11–37; problem of, 4; questions of, 26–27; religious conferral of, 54; scientific method as source of, 32; vesting life with, 18, 26; world lacks, 3

Meaninglessness, 25; accepting, 28–32

Medical care, 101–3

Medical consequences, of founder effect, 88–89

Medical prognosis, and genetic bottleneck, 72–73

Medical research, placebo effects in, 63–67

Medicine, 36, 56, 72; agenda for, from Jewish tradition, 58–62; confusing religious faith with biological ancestry, 98; and free will, 8–9; and Judaism, 37, 39–42, 57, 58; and meaning and transgenes, 42–45; moral obligations of, 82; and religion, 54–55, 69, 99; in science, 99,

100, 101, 103; source of meaning, 97; transgenic, 49–52; see also DNA-based medicine; Genetic medicine

Memes, 29–32, 72–73, 83; free will and, 35, 36; medicine and religion as, 54; religion as parasitic, 28–32; symbiotic, 33–34, 36

Mental illness, 61

Mental states, 61, 62

Mind, 7, 62; and body, 97; studies of, 18–19

Mind/body dichotomy, 60–61, 62–63

Mishnah, 50, 74–75, 77

Missing Moment, The (Pollack), 18–19, 100

Molecular biology, 39, 40–42

Moral authority of medicine, 59–60, 82

Mormon families, 92

Mortality, 1, 18; fact of, 37, 53; religious practice and, 54, 55; of species, 23; see also Death

Moses, 37

Muscular dystrophy, 52

Muslims, 55

Mutations, 5, 24, 47, 78, 80; BRCA1 gene, 92, 93–94, 96–97; in breast cancer, 92

National Institutes of Health, 100

Natural selection, 1, 2, 7, 26, 28–29, 35, 36; without dogma, 22–24; feelings

Randomness, 34, 35
Rationality, 4
Rejection, 4–6, 37
Religion(s), 37, 49; accepting the irrational in, 18–20; confused with biological ancestry, 97–98; medicine and, 54–55, 69, 99; as parasitic meme, 28–32; and science, xii-xiii, 1–6, 7, 11–37, 102
Religious obligations, 17, 71; choosing, 52–54; free will and, 73–74
Retreat, heroism, 59–60
Revelation, 13–15, 21, 26, 34, 36, 37, 39, 105; not reproducible, 15–18
Rosh Hashanah, 73–74

"St. Jude chain letter," 31
Schnabel, Freya, 96
Schoff Memorial Lectures, xii-xiii, 1, 7, 39
Science: accepting the irrational in, 18–20; as authority, 58, 59; at boundary of known and unknown, 12–13, 14; freedom from data of, 53–54; insight in, 16–18; medical interventions, 57; medicine in, 99, 100, 101, 103; memes of, 31–32; moral context of, 99; of one life at a time, 71–103; putting in God's place, 4; and religion, xii-xiii, 1–6, 7, 11–37, 102; religion as source of

meaning for data of, 97–98; and use of DNA information, 100
Science magazine, 26, 28
Scientific American, 63
Scientific discourse, and names of God, 6–7
Scientific fallacy, 48–49
Scientific method, 13, 14, 32, 37
Scientific studies, with no medical purpose, 93–94
Selfish Gene, The (Dawkins), 28, 36
Sexual reproduction, 75
Shem, 87
Siblings, 75, 76
Sickle-cell anemia, 78
Solomon, 87–88
Soloveitchik, Joseph, 59
Soul, 5, 63, 67
Species: common ancestry of, 22–23, 46; finite lifetime of, 23; origins of, 23–2; survival of, 77, 78
Steinsaltz, Adin, 3, 35, 58, 74
Strife of the Spirit, The (Steinsaltz), 35
Survival: differential, 29; of ideas, 32; selection for, 32; of species, 77, 78
Symbiosis, 32–33

Talmud, 39, 55–56, 57, 74; on medicine, 57
Tay-Sachs Disease, 90, 91